UNIVERSITY MATHEMATICAL TEXTS

EDITORS/ALAN JEFFREY AND IAIN T. ADAMSON

ELEMENTARY RINGS AND MODULES

RECENT UNIVERSITY MATHEMATICAL TEXTS

Introduction to Field Theory/I. T. Adamson
An Introduction to Modern Mathematics/Albert Monjallon
Quantum Mechanics/R. A. Newing and J. Cunningham
Probability/J. R. Gray
Series Expansions for Mathematical Physicists/H. Meschkowski
An Introduction to the Theory of Statistics/R. L. Plackett
Mathematical Analysis/G. H. Fullerton
Metric Spaces/C. G. C. Pitts

IAIN T. ADAMSON
SENIOR LECTURER IN MATHEMATICS
UNIVERSITY OF DUNDEE

ELEMENTARY RINGS AND MODULES

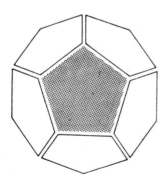

BARNES & NOBLE BOOKS · New York
(a division of Harper & Row, Publishers, Inc.)

Published in the U.S.A., 1972, by
HARPER & ROW, PUBLISHERS, Inc.
Barnes & Noble Import Division

OLIVER AND BOYD
Tweeddale Court
14 High Street
Edinburgh EH1 IYL
A Division of Longman Group Limited

UMT 39
ISBN 06 4900169
First Published 1972
© 1972 Iain T. Adamson
All rights reserved

Printed in Great Britain by
Page Bros (Norwich) Ltd., Norwich

PREFACE

Few students of mathematics nowadays can escape from university without being introduced to vector spaces and abelian groups. These structures are important special cases of modules over rings, so that the study of modules over arbitrary rings appears as a very natural sequel to introductory courses in algebra. This book is intended to provide the basic material for such a study; it is essentially self-contained since it includes a quick introduction to the elementary ideas of abstract algebra.

The first chapter is devoted to the exposition of these ideas and of the fundamental facts about rings and their ideals and homomorphisms, with many illustrative examples.

In Chapter 2, which forms the main part of the book, the reader is introduced first to modules, submodules, factor modules and homomorphisms. This introduction is followed by a description of several important topics—groups of homomorphisms, direct products and sums, tensor products and free, projective and injective modules. This chapter concludes with a section on Artinian and Noetherian modules. Although the language of categories is not used here, categorical ideas have largely determined the presentation.

Chapters 1 and 2 are the more elementary sections of my book *Rings, Modules and Algebras*.

Chapter 3 is concerned exclusively with commutative rings and deals with a number of topics which are fundamental to algebraic number theory and modern algebraic geometry. It begins with a section on unique factorisation domains; this is followed by a discussion of maximal and prime ideals, quotient rings and integral dependence. The book concludes with a section on Dedekind domains.

As in the case of *Rings, Modules and Algebras*, so with this book I am indebted to Kingswood College in the University of Western Australia for giving me a congenial home while I was writing, to Dr Arthur Sands and Dr Hamish Anderson for their work on the manuscript and the proofs, and to my wife for her long forbearance.

IAIN T. ADAMSON

Dundee.
May 1971.

CONTENTS

CHAPTER 1   **RINGS AND IDEALS**

    1   Internal laws of composition   **1**
    2   Semigroups, groups and rings   **7**
    3   Subrings and ideals   **14**
    4   Homomorphisms of rings   **23**
        Exercises 1   **29**

CHAPTER 2   **MODULES**

    5   Modules, submodules and factor modules   **32**
    6   Homomorphisms of modules   **41**
    7   Groups of homomorphisms   **46**
    8   Direct products and sums   **55**
    9   Free, projective and injective modules   **65**
   10 Tensor products   **77**
   11 Artinian and Noetherian modules   **85**
        Exercises 2   **90**

CHAPTER 3   **COMMUTATIVE RINGS**

   12 Unique factorisation domains   **94**
   13 Maximal ideals and prime ideals   **104**
   14 Quotient rings   **108**
   15 Integral dependence   **116**
   16 Dedekind domains   **120**
        Exercises 3   **132**

        Reading list   **134**

        Index   **135**

CHAPTER 1
RINGS AND IDEALS

§1. Internal Laws of Composition

Let E be any set; by an *internal law of composition* or *binary operation* on E we mean a mapping from the Cartesian product $E \times E$ to E. If φ is an internal law of composition on E, the image under φ of an ordered pair (a, b) in $E \times E$ is usually not denoted by $\varphi(a, b)$, which is the notation we would expect from our acquaintance with elementary set theory. The various alternative notations used for $\varphi(a, b)$ include $a + b$ (in which case we call $\varphi(a, b)$ the *sum* of a and b, and refer to φ as an *addition operation*) and ab, $a . b$ or $a \times b$ (when we call $\varphi(a, b)$ the *product* of a and b, and describe φ as a *multiplication operation*). Even when we discuss completely arbitrary internal laws of composition we use a notation reminiscent of these long-familiar ones and write $a \circ b$, $a * b$, $a \wedge b$ or $a \vee b$ instead of $\varphi(a, b)$.

The ordinary addition and multiplication operations are internal laws of composition on the sets $\mathbf{N}$, $\mathbf{Z}$, $\mathbf{Q}$, $\mathbf{R}$, $\mathbf{C}$ of natural numbers, integers, rational numbers, real numbers and complex numbers. Subtraction is an internal law of composition on $\mathbf{Z}$, $\mathbf{Q}$, $\mathbf{R}$, $\mathbf{C}$, and so is division on the sets of non-zero elements of $\mathbf{Q}$, $\mathbf{R}$ and $\mathbf{C}$. If X is any set, composition of mappings from X to itself is an internal law of composition in the set $\mathrm{Map}(X, X)$.

Let E be a set, φ an internal law of composition on E; for each ordered pair (x, y) in $E \times E$ we shall denote $\varphi(x, y)$ by $x \wedge y$. A subset A of E is said to be *closed* or *stable* under φ (or loosely, under $\wedge$) if for every ordered pair (a, b) in $A \times A$ the element $\varphi(a, b) = a \wedge b$ of E actually belongs to the subset A itself. When A is closed under φ we may say that φ *induces* an internal law of composition φ_A on A; φ_A is given by $\varphi_A(a, b) = \varphi(a, b) = a \wedge b$ for all ordered pairs (a, b) in $A \times A$.

Let n be a positive integer, and let I be the interval $[1, n] = \{i \in \mathbf{Z} : 1 \leqslant i \leqslant n\}$. If $(a_i)_{i \in I}$ is a sequence of elements of E we define the *composition* $\wedge_{i \in I} a_i$ of this sequence by induction on n as follows:

(1) If $n = 1$, we set $\wedge_{i \in I} a_i = a_1$;

(2) If $n = k + 1$, let $I' = [1, k]$ and set $\wedge_{i \in I} a_i = (\wedge_{i \in I'} a_i) \wedge a_{k+1}$.

1

When we use the additive notation for φ we write $\sum_{i \in I} a_i$ for the composition of the sequence $(a_i)_{i \in I}$ and call it the *sum* of the sequence; when we use the multiplicative notation we denote the composition by $\prod_{i \in I} a_i$ and call it the *product* of the sequence.

Now suppose that we have two sets E_1, E_2 and internal laws of composition φ_1, φ_2 on E_1, E_2 respectively; write $\varphi_i(x_i, y_i) = x_i \wedge_i y_i$ for every pair (x_i, y_i) in $E_i \times E_i$ $(i = 1, 2)$. A mapping α from E_1 to E_2 is said to be a *homomorphism* (relative to φ_1 and φ_2) if for every pair (x_1, y_1) in $E_1 \times E_1$ we have

$$\alpha(x_1 \wedge_1 y_1) = \alpha(x_1) \wedge_2 \alpha(y_1).$$

If we introduce the mapping $\alpha \times \alpha$ from $E_1 \times E_1$ to $E_2 \times E_2$ by setting $(\alpha \times \alpha)(x_1, y_1) = (\alpha(x_1), \alpha(y_1))$ for each pair (x_1, y_1) in $E_1 \times E_1$, we see that α is a homomorphism if and only if the diagram

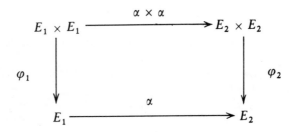

is commutative.

Let α be a homomorphism from E_1 to E_2. If α is surjective we call it an *epimorphism* from E_1 onto E_2; if α is injective we call it a *monomorphism*; and if α is bijective we say that it is an *isomorphism* of E_1 onto E_2. If $E_1 = E_2 = E$ and $\varphi_1 = \varphi_2 = \varphi$, a homomorphism from E to E (relative to φ and φ) is called an *endomorphism* of E (relative to φ); a bijective endomorphism is called an *automorphism* of E. As we have already done in this paragraph we shall usually omit any reference to the laws of composition when they are clear from the context.

We have the following general result concerning homomorphisms.

THEOREM 1.1. *Let α be a homomorphism from E_1 to E_2 relative to internal laws of composition φ_1 and φ_2 on E_1 and E_2 respectively. Then $\operatorname{Im} \alpha = \alpha(E_1)$ is closed under φ_2.*

Proof. Let a_2, b_2 be elements of Im α. Then there are elements a_1, b_1 of E_1 such that $a_2 = \alpha(a_1)$, $b_2 = \alpha(b_1)$. Since α is a homomorphism we have $a_2 \wedge_2 b_2 = \alpha(a_1) \wedge_2 \alpha(b_1) = \alpha(a_1 \wedge_1 b_1)$; so $a_2 \wedge_2 b_2 \in$ Im α, as required.

We shall also use from time to time the results of the following two theorems.

THEOREM 1.2. *Let E_1, E_2, E_3 be sets with internal laws of composition φ_1, φ_2, φ_3 respectively. If α and β are homomorphisms from E_1 to E_2 and E_2 to E_3 respectively, then $\beta\alpha$ is a homomorphism from E_1 to E_3.*

Proof. Let a and b be any two elements of E_1. Then we have $\beta\alpha(a \wedge_1 b) = \beta(\alpha(a \wedge_1 b)) = \beta(\alpha(a) \wedge_2 \alpha(b)) = \beta(\alpha(a)) \wedge_3 \beta(\alpha(b)) = \beta\alpha(a) \wedge_3 \beta\alpha(b)$. Thus $\beta\alpha$ is a homomorphism, as asserted.

THEOREM 1.3. *Let E_1 and E_2 be sets with internal laws of composition φ_1 and φ_2 respectively. If α is an isomorphism of E_1 onto E_2 then α^{-1} is an isomorphism of E_2 onto E_1.*

Proof. Since α is a bijection from E_1 onto E_2, its inverse α^{-1} is a bijection from E_2 onto E_1.

Now let c and d be elements of E_2; let $a = \alpha^{-1}(c)$, $b = \alpha^{-1}(d)$. Then, since α is a homomorphism, we have $\alpha(a \wedge_1 b) = \alpha(a) \wedge_2 \alpha(b) = c \wedge_2 d$ and hence $\alpha^{-1}(c \wedge_2 d) = a \wedge_1 b = \alpha^{-1}(c) \wedge_1 \alpha^{-1}(d)$. Thus α^{-1} is a homomorphism, and hence an isomorphism, as required.

Once again let φ be an internal law of composition on a set E; we shall write $\varphi(a, b) = a \wedge b$ for each pair (a, b) in $E \times E$. We say that φ (or loosely, $\wedge$) is *associative* if for all elements a, b, c of E we have

$$a \wedge (b \wedge c) = (a \wedge b) \wedge c. \qquad [1.1]$$

The importance of this property lies in the fact that it allows us to omit parentheses and to write both sides of [1.1] simply as $a \wedge b \wedge$. From this it follows by an inductive argument that we may omit parentheses in the notation for the composition of arbitrary finite sequences of elements.

We remark that associativity is 'preserved under homomorphisms'; more formally, we have the following theorem, whose proof is a simple exercise.

THEOREM 1.4. *Let α be a homomorphism from E_1 to E_2 relative to* *laws of composition* φ_1 *and* φ_2 *on* E_1 *and* E_2 *respectively. If* φ_1 *is associative then the law of composition induced by* φ_2 *on* Im α *is associative also.*

Keeping the same notation as above, we say that ω (or $\wedge$) is *commutative* if for every ordered pair (a, b) in $E \times E$ we have $a \wedge b = b \wedge a$. Even when φ is not commutative there may nevertheless exist some pairs (a, b) in $E \times E$ for which $a \wedge b = b \wedge a$. If (a, b) is such a pair we say that a *commutes* with b or that a and b *commute*.

We have an obvious analogue of Theorem 1.4.

THEOREM 1.5. *Let* α *be a homomorphism from* E_1 *to* E_2 *relative to laws of composition* φ_1 *and* φ_2 *on* E_1 *and* E_2 *respectively. If* φ_1 *is commutative then the law of composition induced by* φ_2 *on* Im α *is also commutative.*

An element n of E is said to be a *left neutral element* with respect to φ if for every element a of E we have $n \wedge a = a$ and a *right neutral element* if for every element a of E we have $a \wedge n = a$; an element which is both left neutral and right neutral is called simply a *neutral element*. If the law of composition φ is an addition operation a neutral element is usually called a *zero element* and denoted by 0 or z or ζ; if φ is a multiplication operation we refer to a neutral element as an *identity* and denote it by 1 or e or I.

THEOREM 1.6. *Let* φ *be an internal law of composition on a set* E. *If* n_l *and* n_r *are left and right neutral elements respectively with respect to* φ, *then* $n_l = n_r$.

Proof. Since n_l is a left neutral element we have $n_l \wedge n_r = n_r$; but since n_r is a right neutral element we have also $n_l \wedge n_r = n_l$. Hence $n_l = n_r$, as asserted.

COROLLARY. *A set can have at most one neutral element with respect to an internal law of composition.*

We also have a theorem about the 'preservation' of neutral elements under homomorphisms.

THEOREM 1.7. *Let* α *be a homomorphism from* E_1 *to* E_2 *relative to internal laws of composition* φ_1 *and* φ_2 *on* E_1 *and* E_2 *respectively. If* n_1 *is a left (or right) neutral element with respect to* φ_1 *in* E_1 *then*

$\alpha(n_1)$ *is a left (or right) neutral element with respect to the law of composition induced by* φ_2 *on* Im α.

Proof. If a_2 is any element of Im α there is an element a_1 of E_1 such that $\alpha(a_1) = a_2$. Then $\alpha(n_1) \wedge_2 a_2 = \alpha(n_1) \wedge_2 \alpha(a_1) = \alpha(n_1 \wedge_1 a_1) = \alpha(a_1) = a_2$ and $\alpha(n_1)$ is a left neutral element, as asserted.

At this point we introduce a useful convention. If the set E has a neutral element n and $(a_i)_{i \in \phi}$ is an empty family of elements of E we define the composition $\wedge_{i \in \phi} a_i$ to be the neutral element n.

Let E be a set which has a neutral element n with respect to an internal law of composition φ (or $\wedge$). An element a of E is said to be *left-invertible* (with respect to φ) if there exists an element a' of E such that $a' \wedge a = n$; in this situation the element a' is called a *left inverse* of a. The terms *right-invertible* and *right inverse* are defined in the obvious way. An element of E which is both left- and right-invertible is said to be *invertible*. If a is invertible and a' is both a left and a right inverse of a, we say simply that a' is an *inverse* of a; in this case, of course, a is an inverse of a'.

THEOREM 1.8. *Let* φ *be an associative internal law of composition on a set* E, *with respect to which* E *has a neutral element* n. *If an element* a *of* E *has a left inverse* a'_l *and a right inverse* a'_r, *then* $a'_l = a'_r$.

Proof. We have $a'_l \wedge a = n = a \wedge a'_r$. Since φ is associative, $(a'_l \wedge a) \wedge a'_r = a'_l \wedge (a \wedge a'_r)$; hence $n \wedge a'_r = a'_l \wedge n$, i.e. $a'_r = a'_l$, as asserted.

We can also show that inverses are 'preserved under homomorphisms'. It will be enough to state the result: the proof is clear.

THEOREM 1.9. *Let* α *be a homomorphism from* E_1 *to* E_2 *relative to internal laws of composition* φ_1 *and* φ_2 *on* E_1 *and* E_2 *respectively. Suppose* E_1 *has a neutral element* n_1. *If* a'_1 *is a left (or right) inverse of an element* a_1 *of* E_1 *with respect to* φ_1 *then* $\alpha(a'_1)$ *is a left (or right) inverse of* $\alpha(a_1)$ *with respect to* φ_2.

There are two further useful elementary results concerning invertible elements which we include in the following theorem.

THEOREM 1.10. *Let* φ *be an associative internal law of composition on a set* E, *with respect to which* E *has a neutral element* n. *If* a *and* b *are invertible elements of* E *then so is their product* $a \wedge b$. *If, further,* *φ is commutative and $a \wedge b$ is invertible, then a and b are themselves invertible.*

Proof. Let us write φ multiplicatively, abbreviating $a \wedge b$ to ab.

(1) Suppose a and b are invertible, with inverses a' and b' respectively. Then $(ab)(b'a') = a(b(b'a')) = a((bb')a') = a(na') = aa' = n$, and similarly $(b'a')(ab) = n$. Thus ab is invertible.

(2) Now suppose that φ is commutative as well as associative and that the product ab is invertible, with inverse c say. Then we have $(ab)c = n$, whence $a(bc) = n$ and $b(ac) = (ba)c = (ab)c = n$. So a and b are invertible, as asserted.

Let now φ and θ be two internal laws of composition on a set E; we shall write $\varphi(a, b) = a \wedge b$ and $\theta(a, b) = a \vee b$ for every pair (a, b) in $E \times E$. Then we say that φ is *left-distributive* over θ (or that $\wedge$ is left-distributive over $\vee$) if for all elements a, b, c of E we have $a \wedge (b \vee c) = (a \wedge b) \vee (a \wedge c)$. Similarly we say that φ is *right-distributive* over θ if for all elements a, b, c of E we have $(b \vee c) \wedge a = (b \wedge a) \vee (c \wedge a)$. If φ is both left- and right-distributive over θ we say simply that it is *distributive over* θ. Clearly if φ is commutative the three properties coincide.

Once again let φ be an internal law of composition on a set E and write as usual $\varphi(a, b) = a \wedge b$. If R is an equivalence relation on E we say that φ (or $\wedge$) is *compatible* with R if for all elements a, a', b, b' of E such that aRa' and bRb' we also have $(a \wedge b)R(a' \wedge b')$. Suppose that φ is compatible with R, and let η be the canonical surjection from E onto E/R; we shall show how to define an internal law of composition φ_* on E/R such that η is a homomorphism from E to E/R relative to φ and φ_*.

To this end, let C_1, C_2 be elements of E/R; choose elements a_1, a_2 from C_1, C_2 respectively, and define $C_1 \wedge_* C_2 = \eta(a_1 \wedge a_2)$. We must check that the equivalence class so obtained depends only on the classes C_1, C_2 and not on the choice of elements a_1, a_2. But if a'_1, a'_2 are also elements of C_1, C_2 respectively, we have $a_1 R a'_1$ and $a_2 R a'_2$; hence, since $\wedge$ is compatible with R, $(a_1 \wedge a_2)R(a'_1 \wedge a'_2)$ and so $\eta(a_1 \wedge a_2) = \eta(a'_1 \wedge a'_2)$. We say that the law of composition $\wedge_*$ is *induced* by $\wedge$. It is clear that η is a homomorphism from E to E/R relative to $\wedge$ and $\wedge_*$; for if x and y are elements of E they are also elements of the equivalence classes $\eta(x)$ and $\eta(y)$ respectively and so $\eta(x) \wedge_* \eta(y) = \eta(x \wedge y)$ by the definition of $\wedge_*$.

§2. Semigroups, Groups and Rings

Let φ be an internal law of composition on a set E and write as usual $\varphi(a, b) = a \wedge b$. If φ is associative we say that E is a *semigroup* under φ (or, under $\wedge$). Usually, once we have described the law of composition φ with which we are to work, we refer simply to 'the semigroup E' without further explicit mention of φ.

Let E be a semigroup under $\wedge$; let a be an element of E. We proceed to define for each non-zero natural number n the nth *iterate* of a, which we denote by $\wedge^n a$; we set

(1) $\wedge^1 a = a$;
(2) $\wedge^{k+1} a = (\wedge^k a) \wedge a$ for all $k \geqslant 1$.

Thus $\wedge^n a$ is the composition under $\wedge$ of a sequence $(a_i)_{i \in I}$ of elements of E such that Card $I = n$ and $a_i = a$ for every index i in I. If E has a neutral element e with respect to $\wedge$, we further define

(3) $\wedge^0 a = e$,

and if a has an inverse a' relative to $\wedge$ then for every natural number n we set

(4) $\wedge^{-n} a = \wedge^n(a')$.

When $\wedge$ is an addition operation we usually write na instead of $\wedge^n a$ and call it the nth *multiple* of a; when $\wedge$ is a multiplication operation we use the notation a^n and call this the nth *power* of a. Fairly straightforward inductive arguments serve to establish the following results which (at least in the additive and multiplicative notations) look very familiar.

THEOREM 2.1. *Let E be a semigroup under $\wedge$: let a be an element of E. Then for all integers m and n for which the relevant iterates are defined, we have*

$$\wedge^{m+n} a = (\wedge^m a) \wedge (\wedge^n a), \quad \wedge^m(\wedge^n a) = \wedge^{mn} a.$$

If, further, b is an element of E which commutes with a, then

$$\wedge^m(a \wedge b) = (\wedge^m a) \wedge (\wedge^m b).$$

A semigroup is said to be a *group* if it has a neutral element relative to the law of composition and every element has an inverse relative to the law of composition. A group is said to be *abelian* if its law of composition is commutative.

Now let R be a set equipped with an addition operation and a multiplication operation; that is to say, we are given two internal laws of composition on R, one of which is written additively, the other multiplicatively. We say that R is a *ring* under these laws of composition if

(1) R is an abelian group under the addition operation;
(2) R is a semigroup under the multiplication operation;
(3) The multiplication operation is distributive over the addition operation.

We now spell out in detail what is involved in these conditions: R is a ring under the given addition and multiplication operations if

A1. The addition is associative, i.e. for all elements a, b, c of R we have $a + (b + c) = (a + b) + c$.

A2. The addition is commutative, i.e. for all elements a, b of R we have $a + b = b + a$.

A3. There is a neutral element relative to the addition, i.e. an element of R which we call the zero element and denote by 0, such that for every element a of R we have $a + 0 = a = 0 + a$.

A4. Every element of R has an inverse relative to the addition, i.e. for each element a of R there exists an element of R which we call the negative of a and denote by $-a$ such that $a + (-a) = 0 = (-a) + a$.

M1. The multiplication is associative, i.e. for all elements a, b, c of R we have $a(bc) = (ab)c$.

AM. The multiplication is distributive over the addition, i.e. for all elements a, b, c of R we have $a(b + c) = ab + ac$ and $(b + c)a = ba + ca$.

A ring R is called a *commutative ring* if, in addition to the defining properties, it also satisfies

M2. The multiplication is commutative, i.e. for all elements a, b of R we have $ab = ba$.

A ring R is called a *ring with identity* if it satisfies conditions *A1-4*, *M1*, *AM* and, in addition,

M3. There is a neutral element for the multiplication, i.e. an element e of R which we call the identity of R such that for every element a of R we have $ae = a = ea$.

If a, b, c are elements of a ring R such that $a = bc$ we say that a is a *left multiple* of c and a *right multiple* of b; we say also that c is a *right divisor* of a and that b is a *left divisor* of a. A little later we shall see that for every element a of R we have $a0 = 0$; so, according to the definition we have just given, every element of R is a left divisor of zero. It is conventional, however, to make a further condition in this case, and to say that an element b of a ring R is a *left divisor of zero* if and only if it is itself non-zero and there exists a *non-zero* element c of R such that $bc = 0$. *Right divisors of zero* are defined in a similarly restricted way.

A ring R is said to be an *integral domain* if it is commutative (i.e. it satisfies *M2*) and in addition satisfies

M4*. There are no divisors of zero in R, i.e. if a and b are elements of R such that $ab = 0$ then either $a = 0$ or $b = 0$.

A ring R is called a *division ring* if it has at least two elements, satisfies the condition *M3* and in addition

M4. Every non-zero element of R has an inverse relative to multiplication, i.e. for each non-zero element a of R there exists an element a^{-1} of R such that $aa^{-1} = e = a^{-1}a$.

Finally, a commutative division ring is called a *field*; thus a field is a ring with at least two elements satisfying conditions *A1-4*, *M1-4* and *AM*.

Example 1. The set of integers $\mathbf{Z}$ is an integral domain with identity under the ordinary addition and multiplication operations. The only elements of $\mathbf{Z}$ which have multiplicative inverses are 1 and -1; so $\mathbf{Z}$ is not a field under its ordinary addition and multiplication.

Example 2. The set of even integers is an integral domain under the ordinary addition and multiplication operations. There is no identity element and so it does not even make sense to ask whether there are any elements with multiplicative inverses.

Example 3. Let R be any set with an addition operation under which it forms an abelian group. Define a multiplication operation in R by setting $ab = 0$ for every pair of elements a, b of R. Then R is a commutative ring under the given addition and the multiplication we have defined. If R does not consist of the zero element alone there is no identity element and R is not an integral domain.

Example 4. The sets $\mathbf{Q}$, $\mathbf{R}$, $\mathbf{C}$ of rational, real and complex numbers are fields under the ordinary addition and multiplication operations in these sets.

Example 5. Let E be any set, A any ring; consider the set $R = \mathrm{Map}\,(E, A)$ of all mappings from E to A. We define operations of addition and multiplication in R as follows: let α and β be mappings

B

from E to A, and define $\alpha + \beta$ and $\alpha.\beta$ to be the mappings from E to A given by

$$(\alpha + \beta)(x) = \alpha(x) + \beta(x)$$
$$(\alpha.\beta)(x) = \alpha(x)\,\beta(x)$$

for every element x of E. Then R is a ring under these operations. The conditions *A1*, *A2*, *M1* and *AM* for R follow from the same conditions for A; the zero element of R is the mapping ζ from E to A defined by setting $\zeta(x) = 0$ for every element x of E; and if α is any element of R its negative is the mapping $-\alpha$ from E to A given by $(-\alpha)(x) = -(\alpha(x))$ for all elements x of E. If the ring A is commutative then so is R. If A has an identity element e, then the mapping ε from E to A, defined by setting $\varepsilon(x) = e$ for every element x of E, is an identity for R. If A is a field then every element α of R such that $\alpha(x) \neq 0$ for every element x of E has an inverse in R: the inverse is the mapping α^{-1} from E to A given by $\alpha^{-1}(x) = (\alpha(x))^{-1}$ for every element x of E.

Example 6. Let A be an additive abelian group, i.e. a set which forms an abelian group under an internal law of composition which is written additively; let End A be the set of endomorphisms of A. If α and β are elements of End A we define mappings $\alpha + \beta$ and $\alpha\beta$ from A to itself by setting

$$(\alpha + \beta)(a) = \alpha(a) + \beta(a) \quad \text{and} \quad (\alpha\beta)(a) = \alpha(\beta(a)) \qquad [2.1]$$

for all elements a of A. We claim that these mappings are in fact endomorphisms of A. So let a and b be any two elements of A; then we have

$$\begin{aligned}
(\alpha + \beta)(a + b) &= \alpha(a + b) + \beta(a + b) \\
&= (\alpha(a) + \alpha(b)) + (\beta(a) + \beta(b)) \\
&= (\alpha(a) + \beta(a)) + (\alpha(b) + \beta(b)) \\
&= (\alpha + \beta)(a) + (\alpha + \beta)(b)
\end{aligned}$$

and

$$\begin{aligned}
(\alpha\beta)(a + b) = \alpha(\beta(a + b)) &= \alpha(\beta(a) + \beta(b)) \\
&= \alpha(\beta(a)) + \alpha(\beta(b)) = (\alpha\beta)(a) + (\alpha\beta)(b).
\end{aligned}$$

Thus the prescriptions [2.1] actually define internal laws of composition on End A. We shall now show that under these laws End A

is a ring with identity. The conditions *A1* and *A2* for End *A* follow from the associativity and commutativity of the group operation in *A*; the mapping ζ from *A* to *A* given by $\zeta(a) = 0$ for all elements *a* of *A* is easily seen to be an endomorphism of *A* and to be a zero element for End *A*; if α is any endomorphism of *A* then the mapping $-\alpha$ from *A* to *A* given by $(-\alpha)(a) = -(\alpha(a))$ for all elements *a* of *A* is quickly shown to be an endomorphism of *A* and clearly $\alpha + (-\alpha) = \zeta$. Condition *M1* is a basic property of composition of mappings. To show that condition *AM* holds in End *A*, let α, β, γ be endomorphisms of *A* and let *a* be any element of *A*. Then

$$((\alpha + \beta)\gamma)(a) = (\alpha + \beta)(\gamma(a)) = \alpha(\gamma(a)) + \beta(\gamma(a)) = \alpha\gamma(a) + \beta\gamma(a)$$
$$= (\alpha\gamma + \beta\gamma)(a),$$

whence $(\alpha + \beta)\gamma = \alpha\gamma + \beta\gamma$; on the other hand,

$$(\gamma(\alpha + \beta))(a) = \gamma((\alpha + \beta)(a)) = \gamma(\alpha(a) + \beta(a)),$$

while

$$(\gamma\alpha + \gamma\beta)(a) = \gamma\alpha(a) + \gamma\beta(a) = \gamma(\alpha(a)) + \gamma(\beta(a)) = \gamma(\alpha(a) + \beta(a)),$$

so that $\gamma(\alpha + \beta) = \gamma\alpha + \gamma\beta$. Thus End *A* satisfies *AM*. The identity map I_A of *A* belongs to End *A* and is an identity element. We call End *A* the *ring of endomorphisms* of *A*.

Example 7. Let *A* be any ring; then for each positive integer *n* the set $M_n(A)$ of $n \times n$ matrices with coefficients in *A* is a ring under ordinary matrix addition and multiplication. In general the ring $M_n(A)$ is not commutative; it has an identity if *A* does.

Example 8. Let **H** be the four-dimensional vector space over the real number field **R** consisting of all ordered quadruples (a, b, c, d) of real numbers with the vector space addition and scalar multiplication given by

$$(a, b, c, d) + (a', b', c', d') = (a + a', b + b', c + c', d + d'),$$

$$r(a, b, c, d) = (ra, rb, rc, rd)$$

for all real numbers $a, a', b, b', c, c', d, d', r$. We denote the natural basis vectors $(1, 0, 0, 0)$, $(0, 1, 0, 0)$, $(0, 0, 1, 0)$, $(0, 0, 0, 1)$ by $\mathbf{e}, \mathbf{i}, \mathbf{j}, \mathbf{k}$ respectively; so we may write $(a, b, c, d) = a\mathbf{e} + b\mathbf{i} + c\mathbf{j} + d\mathbf{k}$. We now define a multiplication operation in **H** by prescribing the products of the basis vectors according to the table

	e	i	j	k
e	e	i	j	k
i	i	$-e$	k	$-j$
j	j	$-k$	$-e$	i
k	k	j	$-i$	$-e$

and then defining the product of two arbitrary vectors in the natural way:

$$(a\mathbf{e} + b\mathbf{i} + c\mathbf{j} + d\mathbf{k})(a'\mathbf{e} + b'\mathbf{i} + c'\mathbf{j} + d'\mathbf{k})$$
$$= (aa')\,\mathbf{e}^2 + (ab')\,\mathbf{ei} + (ac')\,\mathbf{ej} + (ad')\,\mathbf{ek}$$
$$+ (ba')\,\mathbf{ie} + (bb')\,\mathbf{i}^2 + (bc')\,\mathbf{ij} + (bd')\,\mathbf{ik}$$
$$+ (ca')\,\mathbf{je} + (cb')\,\mathbf{ji} + (cc')\,\mathbf{j}^2 + (cd')\,\mathbf{jk}$$
$$+ (da')\,\mathbf{ke} + (db')\,\mathbf{ki} + (dc')\,\mathbf{kj} + (dd')\,\mathbf{k}^2. \qquad [2.3]$$

(We say that we 'extend by distributivity'.) Then it can be verified that under the addition and multiplication defined $\mathbf{H}$ is a non-commutative ring with identity, the identity being the basis vector $\mathbf{e}$. We claim that $\mathbf{H}$ is in fact a division ring. Let $\mathbf{q} = a\mathbf{e} + b\mathbf{i} + c\mathbf{j} + d\mathbf{k}$ be a non-zero element of $\mathbf{H}$; then a, b, c, d are not all zero and hence $a^2 + b^2 + c^2 + d^2 \neq 0$. Let $\mathbf{q}^* = a\mathbf{e} - b\mathbf{i} - c\mathbf{j} - d\mathbf{k}$; then, using [2.3] above we readily compute $\mathbf{qq}^* = (a^2 + b^2 + c^2 + d^2)\,\mathbf{e}$. Thus $(a^2 + b^2 + c^2 + d^2)^{-1}\mathbf{q}^*$ is a multiplicative inverse for $\mathbf{q}$. So $\mathbf{H}$ is a non-commutative division ring; we call it the ring of *real quaternions*.

We now mention some very elementary properties of rings. First of all it follows from Theorem 1.6 that the zero element of a ring and the identity element of a ring with identity are unique, and from Theorem 1.8 that the negatives of all the elements of a ring and the multiplicative inverses of all the elements which are invertible with respect to multiplication are unique.

Next, we show that in a ring subtraction is uniquely possible. So let a and b be elements of a ring R; to subtract b from a we must find an element x of R such that $a = x + b$. Clearly $a + (-b)$ satisfies this requirement; and if x_1 is any element of R such that $a = x_1 + b$ then, on adding $-b$ to both sides and applying $A1$ and $A4$, we have $x_1 = a + (-b)$. We shall write $a - b$ instead of $a + (-b)$.

Now let a be any element of R; we shall show that $0a = 0$. First we remark that, by $A3$, $0 + 0 = 0$; whence $(0 + 0)a = 0a$. By AM this yields $0a + 0a = 0a$; adding $-0a$ to both sides and using $A1$ and $A4$ we obtain $0a = 0$. Similarly we can show that $a0 = 0$.

The last result shows that if the ring R has more than one element then the zero element of R cannot be an identity; for if e is an identity of R and a is any non-zero element we have $ea = a \neq 0$, while we have just seen that $0a = 0$. We deduce also that if R has more than one element and has an identity e then 0 is not invertible with respect to multiplication; for if a were a multiplicative inverse of 0 we would have $e = a0 = 0$, which is a contradiction.

Next let a and b be any two elements of R; we claim that we have $a(-b) = (-a)b = -(ab)$ and that $(-a)(-b) = ab$. Since $b + (-b) = 0$ we deduce that $a(b + (-b)) = a0$, which is zero by what we have just proved. Using AM we have $ab + a(-b) = 0$; so $a(-b)$ is an additive inverse for ab. Since additive inverses are unique we deduce that $a(-b) = -(ab)$. The other results follow by similar arguments.

Suppose now that R has an identity e, and let U be the set of elements of R which have multiplicative inverses; these elements are called the *units* of the ring R. We shall show that U forms a group under the multiplication operation in R. First we must show that U is closed under this operation; but this is an immediate consequence of Theorem 1.10. Thus the multiplication in the ring R induces an internal law of composition on U. This law of composition is clearly associative, since the multiplication in R is associative $(M1)$. The identity of R is its own multiplicative inverse and hence belongs to U, and it is of course an identity for U. Finally, by definition, every element of U has an inverse in R; but if $a \in U$, the inverse a^{-1} of a also belongs to U, since it is invertible (with inverse a). Thus U forms a group under the internal law of composition induced on U by the multiplication in R. We call U the *group of units* of R. When D is a division ring, the group of units consists of all the non-zero elements; in this case we call it the *multiplicative group* of tl e division ring D and denote it by D^*.

We remark also that in a division ring there are no divisors of zero. For suppose a and b are elements of a division ring D with identity e such that $ab = 0$; if a is non-zero, it has an inverse a^{-1} in D and hence

$$0 = a^{-1}0 = a^{-1}(ab) = (a^{-1}a)b = eb = b.$$

Finally, let R be any ring and as usual let the sum and product of any two elements a and b of R be denoted by $a + b$ and ab respectively. We define a new internal law of composition $*$ in R by setting

$$a * b = ba$$

for all elements a, b of R. Then the set of elements of R, equipped with the original addition operation and this new law of composition as multiplication, is a ring which we call the *opposite ring* of R and denote by R^{op}.

§3. Subrings and Ideals

Let S be a subset of a ring R. If S is closed under the addition and multiplication operations in R, then these operations induce internal laws of composition on S. So it makes sense to ask whether S forms a ring under these induced laws of composition; to answer this question it looks at first sight as if we would have to check whether the defining conditions *A1-4, M1* and *AM* of §2 are all satisfied. But a moment's thought should convince the reader that conditions *A1, A2, M1, AM* are automatically satisfied in S.

Suppose now that *A3* is satisfied in S, i.e. suppose that there is a zero element in S, which we denote for a moment by 0_S, such that $s + 0_S = s$ for every element s of S. We claim that 0_S must in fact be the zero element 0 of R; for if a is any element of R we have $a = a + (-(0_S) + 0_S) = (a + (-0_S)) + 0_S$ and hence $a + 0_S = ((a + (-0_S)) + 0_S) + 0_S = (a + (-0_S)) + (0_S + 0_S) = (a + (-0_S)) + 0_S = a$. This shows that 0_S is a neutral element for addition in R, and since neutral elements are unique it follows that $0_S = 0$. Conversely, of course, if 0 belongs to S then it is certainly a zero element of S. Thus *A3* is satisfied in S if and only if 0 belongs to S.

Next suppose that 0 belongs to S and that *A4* is satisfied in S, i.e. suppose that for every element s of S there exists an element s' of S such that $s + s' = 0$. Then s' is an additive inverse of s in S, and hence in R. Since additive inverses are unique it follows that $s' = -s$; so the negative of each element of S in R actually belongs to S. Conversely, if the negative of each element of S belongs to S then *A4* is satisfied in S.

We say that a subset S of a ring R is a *subring* of R if it is closed under the addition and multiplication operations of R and forms a ring under the induced internal laws of composition. We may thus sum up the preceding discussion in the following theorem.

THEOREM 3.1. *A subset S of a ring R is a subring if and only if the following conditions are satisfied*: (1) *S is closed under the addition and multiplication operations of R*, (2) *the zero element of R belongs to S*, (3) *for every element s of S its negative* $-s$ *belongs to S*.

There is also another criterion for a subset to be a subring, which is often more convenient to apply in practice.

THEOREM 3.2. *A non-empty subset S of a ring R is a subring if and only if for every pair of elements s and t of S the elements s* $-$ *t and st belong to S.*

Proof. (1) Suppose S is a subring of R and let s and t be any elements of S.

By condition (3) of Theorem $3.1, -t \in S$. Hence, by condition (1) of 3.1, $s + (-t) = s - t$ belongs to S. Also by condition (1) of 3.1, $st \in S$.

(2) Conversely, suppose that for every pair of elements s, t in S we have $s - t \in S$ and $st \in S$.

The second condition shows that S is closed under the multiplication operation in R.

If s is any element of S it follows from the first condition that $s - s = 0$ belongs to S.

Next, if s is any element of S then since $0 \in S$ and $s \in S$ we deduce that $0 - s = -s$ belongs to S.

Finally, if s and t are any elements of S then $s \in S$ and $-t \in S$; so we have $s - (-t) = s + t \in S$. So S is closed under the addition operation in R.

It now follows from Theorem 3.1 that S is a subring of R.

Suppose that S is a subring of a ring R. It is natural to ask whether S inherits from R any of the additional properties which R may enjoy. It is clear that if R is a commutative ring (i.e. satisfies $M2$) then S is also a commutative ring; similarly if R has no divisors of zero then S has no divisors of zero either. But if R is a division ring we cannot deduce that S is a division ring: for although every non-zero element of S certainly has a multiplicative inverse in R, we cannot assert in general that these inverses belong to S. If R has an identity element e and e belongs to S then it is clearly an identity for S. But R may have an identity while S does not and *vice versa*.

A subring A of a ring R is called a *left ideal* if for every element r of R and every element a of A the product ra belongs to A. Similarly, a subring B of R is called a *right ideal* if for every element r of R and every element b of B the product br belongs to B. A subring which is both a left ideal and a right ideal is called a *two-sided ideal* or simply an *ideal*. A simple adaptation of Theorem 3.2 yields the following criterion for a subset of a ring to be an ideal.

THEOREM 3.3. *A non-empty subset A of a ring R is a left (right) ideal if and only if for all elements a, b of A and r of R the elements $a - b$ and ra (ar) belong to A.*

From this we deduce at once a useful consequence.

COROLLARY. *The intersection of any non-empty collection of left ideals (right ideals, two-sided ideals) of a ring R is a left ideal (right ideal, two-sided ideal) of R.*

The ring R itself is clearly a two-sided ideal of R; the ideals of R (left, right or two-sided) distinct from R itself are called *proper ideals* of R. If R does not consist of the zero element alone then R has at least one proper ideal, for the set $\{0\}$ consisting of the zero element alone is clearly a two-sided ideal of R; we call $\{0\}$ the *zero ideal* of R.

Example 1. Let m be any integer. Then the set of all multiples (positive, negative and zero) of m is a two-sided ideal in the ring of integers $\mathbf{Z}$. We denote this ideal by (m).

We now show that, conversely, every ideal of $\mathbf{Z}$ is of this form. Let I be any ideal of $\mathbf{Z}$. If I is the zero ideal then $I = (0)$. If I is not the zero ideal, then I contains non-zero integers, and hence positive integers (for if z is a non-zero integer in I, $0 - z = -z$ also belongs to I and certainly one of the integers z, $-z$ is positive). Let m be the least positive integer in I; we claim that $I = (m)$. Certainly since I is an ideal all the multiples of m belong to I, i.e. $(m) \subseteq I$. On the other hand, if n is any integer in I we may write $n = qm + r$ where q and r are integers and $0 \leqslant r < m$; since $n \in I$ and $qm \in I$ it follows that $n - qm = r \in I$. But m is the least positive integer in I and $0 \leqslant r < m$; hence $r = 0$ and $n = qm$. Thus $I \subseteq (m)$ and so, finally, $I = (m)$ as asserted.

Example 2. Let E be any set, A any ring; and let R be the ring of mappings from E to A (see §2, Example 5). If x is an element of E, then the set $I(x)$ consisting of all mappings α from E to A such that $\alpha(x) = 0$ is a two-sided ideal of R.

Example 3. Let A be any ring, $R = M_n(A)$ the ring of $n \times n$ matrices with elements in A. Let L_i be the subset of R consisting of matrices which have zero elements in all columns except possibly the ith; then L_i is a left ideal of R. Similarly the subset R_j of R consisting of matrices with zero elements in all rows except possibly the jth is a right ideal of R.

We remark that if R is a ring with identity and I is a proper (left, right or two-sided) ideal of R, then I cannot contain any unit of R. Suppose, for example, that I is a proper left ideal of R and u is a unit of R which belongs to I; then, since $u^{-1} \in R$ and $u \in I$ it follows that $u^{-1}u = e$ belongs to I, and hence if a is any element of R we deduce that $ae = a$ belongs to I; thus $I = R$, contradicting the hypothesis that I is a proper ideal. This discussion shows in particular that if R is a division ring the only ideals (left, right or two-sided) of R are R itself and the zero ideal.

Let X be any subset of a ring R. If E is the set of left ideals of R which include the subset X, then E is certainly non-empty, since R itself belongs to E. According to the Corollary of Theorem 3.3, the intersection $l(X)$ of all the left ideals in the set E is also a left ideal of R; $l(X)$ certainly includes X, and every left ideal which includes X also includes $l(X)$. So $l(X)$ is the smallest left ideal of R which includes X; we call $l(X)$ the left ideal *generated* by X. The right ideal $r(X)$ and the two-sided ideal $t(X)$ of R generated by X are defined in the obvious way; they are respectively the smallest right ideal and the smallest two-sided ideal of R which include X. Since a two-sided ideal is of course both a left and a right ideal, we have $l(X) \subseteq t(X)$ and $r(X) \subseteq t(X)$. If the subset X consists of a single element x of R, we write $l(x)$, $r(x)$, $t(x)$ instead of $l(\{x\})$, $r(\{x\})$, $t(\{x\})$, and we call these ideals the *principal* left, right and two-sided ideals generated by the element x. The next theorem gives a description of the elements of $t(X)$.

THEOREM 3.4. *Let X be a subset of a ring R. Then the two-sided ideal $t(X)$ of R generated by X consists precisely of those elements of R which can be expressed in the form*

$$\sum_{i \in I} n_i x_i + \sum_{j \in J} r_j x_j + \sum_{k \in K} x_k r'_k + \sum_{l \in L} s_l x_l s'_l \qquad [3.1]$$

where I, J, K, L are finite sets (possibly empty), $(x_i)_{i \in I}$, $(x_j)_{j \in J}$, $(x_k)_{k \in K}$ and $(x_l)_{l \in L}$ are families of elements of X, $(n_i)_{i \in I}$ is a family of integers, $(r_j)_{j \in J}$, $(r'_k)_{k \in K}$, $(s_l)_{l \in L}$ and $(s'_l)_{l \in L}$ are families of elements of R, and the elements $n_i x_i$ are the multiples defined in §2.

Proof. Let T be the set of all elements of R which are expressible in the form [3.1].

Then clearly T is included in every two-sided ideal of R which includes X; hence $T \subseteq t(X)$.

But it is easy to check, using the analogue of Theorem 3.3 for two-sided ideals, that T is itself a two-sided ideal of R. Since T includes X, it follows that $T \supseteq t(X)$.

Hence $T = t(X)$, as asserted.

COROLLARY 1. *Let x be an element of a ring R. Then the principal two-sided ideal t(x) generated by x consists of all the elements of R which can be expressed in the form*

$$nx + rx + xr' + \sum_{i \in L} r_i x r'_i \qquad [3.2]$$

where L is a finite set, n is an integer, r and r' are elements of R and $(r_i)_{i \in L}$ and $(r'_i)_{i \in L}$ are families of elements of R.

Similar descriptions can be given for the left and right ideals of R generated by a subset X or an element x of R. For example it is easy to establish that the left ideal $l(X)$ generated by X consists of all elements of R of the form

$$\sum_{i \in I} n_i x_i + \sum_{j \in J} r_j x_j \qquad [3.3]$$

where I and J are finite sets, $(x_i)_{i \in I}$ and $(x_j)_{j \in J}$ are families of elements of X, $(n_i)_{i \in I}$ is a family of integers and $(r_j)_{j \in J}$ is a family of elements of R. In the same way we deduce that the principal right ideal generated by the element x of R is made up of all elements of the form

$$nx + xr \qquad [3.4]$$

where n is an integer and r is an element of R.

When the ring R has an identity element e, it is easily established by an inductive argument that for every element x of R and every integer n we have

$$nx = (ne)\,x = x(ne).$$

Thus when R has an identity the sums $\sum_{i \in I} n_i x_i$ may be dropped from [3.1] and [3.3] and the term nx may be dropped from [3.2] and [3.4], since they may be absorbed in the other terms of the various expressions.

Combining this last remark with Corollary 1 and our descriptions of the left and right ideals generated by an element x of R we have the following useful consequence.

COROLLARY 2. *Let R be a ring with identity, u a unit of R. If x is any element of R then $l(x) = l(ux)$, $r(x) = r(xu)$, $t(x) = t(ux) = t(xu)$.*

If A is a left ideal of a ring R and X is a subset of A such that $l(X) = A$ we say that X is a *generating system* for the left ideal A; if A has a finite generating system we say that A is a *finitely generated* left ideal. In particular, of course, every principal left ideal is finitely generated. The notions of generating system and of finitely generated ideals are defined in the obvious way for right and two-sided ideals.

Let E_* be the set of non-zero left ideals in R. Then the inclusion relation is a relation of order in E_*; if L is a minimal element of E_* under this relation it is called a *minimal left ideal* of R. We similarly define minimal right ideals and minimal two-sided ideals. We remark that a ring R need not have any minimal left, right or two-sided ideals.

Similarly, let E^* be the set of proper left ideals in R. Again the inclusion relation is a relation of order in E^*; a maximal element of E^* under this relation is called a *maximal left ideal* of R. Maximal right ideals and maximal two-sided ideals are defined in the obvious way. If the ring R has no identity element then R need not have any maximal left, right or two-sided ideals; but if R has an identity we deduce as a special case of the next theorem that it must have maximal ideals.

THEOREM 3.5. *Let R be a ring with identity. If A is any proper left ideal of R there exists a maximal left ideal of R which includes A.*

Proof. Let E be the set of proper left ideals which include A; E is certainly non-empty, since A itself belongs to E. The inclusion relation is a relation of order on E; we shall show that E is inductively ordered by this relation.

So let E' be any totally ordered subset of E; we have to show that E' has a least upper bound in E. Let L be the union of all the left ideals in the set E'; we claim that L is a proper left ideal of R which includes A. To this end, let x_1, x_2 be any two elements of L, r any element of R. Then there are left ideals L_1, L_2 say, in the set E' such that $x_1 \in L_1$, $x_2 \in L_2$. Since E' is totally ordered, either $L_1 \subseteq L_2$ or $L_2 \subseteq L_1$, say $L_2 \subseteq L_1$; then x_1 and x_2 both belong to L_1. It follows (by Theorem 3.3) that $x_1 - x_2 \in L_1$ and $rx_1 \in L_1$; hence $x_1 - x_2 \in L$ and $rx_1 \in L$. Thus L is a left ideal. To show that L is a proper left ideal, we have only to remark that, since the ideals in the set E' are all proper, none of them contains the identity of R; hence L does not contain the identity of R; since L clearly includes A it follows that L belongs to E. It is obvious that L is the least upper bound of E'.

The desired result now follows at once by an application of Zorn's Lemma.

COROLLARY. *Every ring with identity has at least one maximal left ideal.*

Now let A and B be left ideals of a ring R. We define the *sum* of A and B to be the subset $A + B$ of R consisting of all elements x of R which can be expressed in the form $x = a + b$ where $a \in A$ and $b \in B$. It is easy to show that $A + B$ is a left ideal of R. First suppose that x_1 and x_2 are elements of $A + B$; then there exist elements a_1, a_2 of A and b_1, b_2 of B such that $x_1 = a_1 + b_1$ and $x_2 = a_2 + b_2$, and hence

$$x_1 - x_2 = (a_1 + b_1) - (a_2 + b_2) = (a_1 - a_2) + (b_1 - b_2).$$

Since $a_1 - a_2 \in A$ and $b_1 - b_2 \in B$ (because A and B are ideals), it follows that $x_1 - x_2 \in A + B$. Next let $x = a + b$ (with a in A and b in B) be any element of $A + B$, r any element of R; then

$$rx = r(a + b) = ra + rb,$$

and so $rx \in A + B$, since $ra \in A$ and $rb \in B$. Thus, by Theorem 3.3, $A + B$ is a left ideal of R. The reader may like to verify that the sum $A + B$ which we have just defined coincides with the left ideal generated by the union $A \cup B$. In the same way we define the sum of two right ideals and two two-sided ideals and show that these sums are respectively right and two-sided ideals.

Again let A and B be left ideals of R. We define the *product* AB of A and B to be the subset of R consisting of all elements which can be expressed in the form $\sum_{i \in I} a_i b_i$ where I is a finite set and $(a_i)_{i \in I}$,

$(b_i)_{i \in I}$ are families of elements of A and B respectively. Using Theorem 3.3 again we can easily show that AB is a left ideal of R. In the same way we can define the product of two right ideals or of two two-sided ideals and show that these products are ideals of the same kind. Having defined the product of two ideals we may apply the procedure described in §2 to define the powers A^n of an ideal A for all natural numbers n: we check easily that A^0 is the ring R itself, which is a neutral element for multiplication of ideals, and that for each non-zero natural number n the nth power A^n consists of all elements of R which can be expressed in the form $\sum_{i \in I} a_{i1} a_{i2} \ldots a_{in}$, where I is a finite set and $(a_{i1})_{i \in I}, \ldots, (a_{in})_{i \in I}$ are families of elements of A.

The following properties of addition and multiplication of ideals in a ring are all easy consequences of the definitions.

THEOREM 3.6. *Let A, B, C be ideals (left, right or two-sided) of a ring R. Then*

A1. $A + (B + C) = (A + B) + C$, M1. $A(BC) = (AB)C$,
A2. $A + B = B + A$,
A3. $A + \{0\} = A$,
AM. $A(B + C) = AB + AC$ and $(A + B)C = AC + BC$.

It follows at once from the definition of ideal multiplication that if A and B are left ideals then $AB \subseteq B$; similarly if A and B are right ideals we have $AB \subseteq A$. Hence, of course, if A and B are two-sided ideals we have $AB \subseteq A$ and $AB \subseteq B$; so $AB \subseteq A \cap B$.

Let R be a ring, I a two-sided ideal of R. Consider the relation defined on R by setting $x \equiv y$ (mod. I) if and only if $x - y \in I$. We claim that this relation, which we read 'x is *congruent* to y modulo I', is an equivalence relation on R. First let x be any element of R; then $x - x = 0$ belongs to I since I is a subring, i.e. $x \equiv x$ (mod. I). Next let x and y be elements of R such that $x \equiv y$ (mod. I); then $x - y \in I$ and so, since I is a subring, $-(x - y) = y - x$ belongs to I, i.e. $y \equiv x$ (mod. I). Finally let x, y, z be elements of R such that $x \equiv y$ (mod. I) and $y \equiv z$ (mod. I). Then $x - y \in I$ and $y - z \in I$; so $(x - y) + (y - z) = x - z$ belongs to I, i.e. $x \equiv z$ (mod. I). Hence the relation is indeed an equivalence relation.

We now show that the addition and multiplication operations in R are compatible with this relation. So suppose $x \equiv x'$ (mod. I) and $y \equiv y'$ (mod. I); then $x - x' \in I$, $y - y' \in I$ and hence we have

$(x - x') + (y - y') = (x + y) - (x' + y') \in I;$ so $x + y \equiv x' + y'$ (mod. I) and the addition in R is compatible with the relation. To prove that the multiplication is compatible we remark that

$$xy - x'y' = xy - xy' + xy' - x'y' = x(y - y') + (x - x')y'.$$

Now $y - y' \in I$ and $x - x' \in I$; and hence, since I is a two-sided ideal, $x(y - y') \in I$ and $(x - x') y' \in I$; it follows that $xy - x'y' \in I$, i.e. $xy \equiv x'y'$ (mod. I).

The quotient set of R with respect to the equivalence relation we have been discussing is denoted by R/I and the equivalence classes are called the *residue classes* modulo the two-sided ideal I; as usual we denote the canonical surjection of R onto R/I by η. According to §1 the addition and multiplication operations in R induce internal laws of composition in R/I as follows: if C_1, C_2 are residue classes modulo I and a_1, a_2 are elements of R in C_1, C_2 respectively, then

$$C_1 + C_2 = \eta(a_1 + a_2) \text{ and } C_1 C_2 = \eta(a_1 a_2). \qquad [3.5]$$

(The compatibility of the operations in R with the equivalence relation assures us that $C_1 + C_2$ and $C_1 C_2$ depend only on C_1 and C_2, not on the choices of a_1 and a_2.) It is now an easy matter to verify that R/I is a ring under the laws of composition induced on it by those in R. The conditions *A1, A2, M1, AM* for R/I all follow directly from the definitions [3.5] and the corresponding conditions for R; the residue class of 0 (which is, of course, the ideal I itself) is a zero element for R/I. Finally, if C is a residue class modulo I, let a be any element of C; then the residue class $\eta(-a)$ containing $-a$ is an additive inverse for C. We call R/I the *residue class ring* of R modulo I. It is clear that if R is a commutative ring then so is R/I, that if R has an identity element e then the residue class $\eta(e)$ is an identity in R/I, and that if the element a of R is invertible in R with inverse a^{-1} then the residue class $\eta(a)$ is invertible in R/I with inverse $\eta(a^{-1})$. But there may be divisors of zero in R/I even when there are none in R itself.

Example 4. Let m be any non-zero integer; let (m) be the two-sided ideal of $\mathbf{Z}$ consisting of all the multiples of m. If a and b are integers we see at once that a is congruent to b modulo the ideal (m) if and only if a is congruent to b modulo the integer m. Since every integer is congruent modulo m to exactly one of the integers $0, 1, \ldots ,$ $m - 1$, it follows that the residue class ring $\mathbf{Z}/(m)$ consists of the m

residue classes $C_0, C_1, \ldots, C_{m-1}$ where, for $i = 0, 1, \ldots, m-1$, C_i consists of all the integers of the form $i + km$ (with k in $\mathbf{Z}$). The residue class C_0 is the zero element of $\mathbf{Z}/(m)$ and the residue class C_1 is the identity. It follows easily from the definition of addition and multiplication in $\mathbf{Z}/(m)$ that if C_a and C_b are any two of the m residue classes then

$$C_a + C_b = C_s \text{ and } C_a C_b = C_p,$$

where s and p are the remainders $(0 \leqslant s, p < m)$ when $a + b$ and ab respectively are divided by m. We notice in particular that if $m = ab$, where $1 < a, b < m$ then $C_a C_b = C_0$; thus if m is not a prime number, $\mathbf{Z}/(m)$ has divisors of zero, even though $\mathbf{Z}$ does not.

Suppose now that p is a prime number; we claim that the residue class ring $\mathbf{Z}/(p)$ is a field. All that remains to be shown is that every non-zero residue class has a multiplicative inverse in $\mathbf{Z}/(p)$. So let C_a be a non-zero element of $\mathbf{Z}/(p)$; then $0 < a < p$ and so, since p is a prime number, the highest common factor of a and p is 1. Hence there exist integers b and q such that $0 < b < p$ and $ab + pq = 1$. Then we have $C_a C_b = C_1$, i.e. C_b is a multiplicative inverse for C_a.

§4. Homomorphisms of Rings

Let R and S be rings. A mapping α from R to S is called a *ring homomorphism* (or simply a homomorphism) if for all elements a, b of R we have

$$\alpha(a + b) = \alpha(a) + \alpha(b) \text{ and } \alpha(ab) = \alpha(a)\,\alpha(b).$$

The other terms, monomorphism epimorphism, ..., defined in §1 are all used in an obvious way.

Example 1. Let R and S be rings, ζ the mapping from R to S defined by setting $\zeta(a) = 0$ for every element a of R. Then ζ is a homomorphism from R to S, which we call the *zero homomorphism*.

Example 2. Let R be a subring of a ring S, ι the canonical injection of R into S. Then ι is a monomorphism from R to S which we call the *inclusion monomorphism*. If $R = S$ and I_R is the identity mapping from R onto itself, I_R is an automorphism, which we call the *identity automorphism* of R.

Example 3. Let R be any ring, I a two-sided ideal of R and R/I the residue class ring. The addition and multiplication operations in R/I are defined in such a way that the canonical surjection η from R onto R/I is a homomorphism. From now on we shall call η the *canonical epimorphism* from R onto R/I.

Gathering together the general results of Theorems 1.1, 1.5, 1.7, 1.9 and 3.2 we can at once make the following statement about ring homomorphisms.

THEOREM 4.1. *Let R and S be rings, α a homomorphism from R to S. Then* Im $\alpha = \alpha(R)$ *is a subring of S. If 0 is the zero element of R then $\alpha(0)$ is the zero element of S; if a is any element of R then $\alpha(-a)$ is the additive inverse of $\alpha(a)$; if R is commutative, so is* Im α; *if R has an identity element e then $\alpha(e)$ is an identity element of* Im α; *if an element a of R has a multiplicative inverse, then $\alpha(a)$ has a multiplicative inverse in* Im α, *namely $\alpha(a^{-1})$.*

Theorems 1.2 and 1.3 show that the composition of two ring homomorphisms is a homomorphism and that the inverse mapping of a ring isomorphism is also an isomorphism.

The *kernel* of a homomorphism α from a ring R to a ring S is the set of elements of R which are mapped by α onto the zero element of S; we denote the kernel of α by Ker α. Thus

$$\text{Ker } \alpha = \{x \in R \,|\, \alpha(x) = 0\} = \alpha^{-1}(0).$$

The kernel of α is clearly non-empty since, as we mentioned in Theorem 4.1, $\alpha(0) = 0$; so the zero element of R belongs to the kernel of every homomorphism of R. Our next theorem shows that if the kernel of a homomorphism contains no other element then the homomorphism is injective.

THEOREM 4.2. *Let α be a homomorphism from a ring R to a ring S. Then α is a monomorphism if and only if* Ker $\alpha = \{0\}$.

Proof. (1) Suppose α is a monomorphism. Then for every element s of lm α the set $\alpha^{-1}(s)$ consists of a single element. In particular Ker $\alpha = \alpha^{-1}(0)$ consists of a single element; and since $\alpha(0) = 0$, this single element is the zero element of R.

(2) Conversely, suppose Ker $\alpha = \{0\}$. We shall show that α is a monomorphism. So let a and b be elements of R such that $\alpha(a) = \alpha(b)$. Since α is a homomorphism, we deduce that $\alpha(a - b) = \alpha(a + (-b))$

$= \alpha(a) + \alpha(-b) = \alpha(a) - \alpha(b) = 0$; so $a - b \in \text{Ker } \alpha$ and hence $a - b = 0$, i.e. $a = b$.

Now we come to the fundamental theorem on homomorphisms of rings.

THEOREM 4.3. (*First homomorphism theorem*). *Let* α *be a homomorphism from a ring* R *to a ring* S *with kernel* K. *Then* K *is a two-sided ideal of* R *and if* η *is the canonical epimorphism from* R *to* R/K, *there exists a unique monomorphism* α_* *from* R/K *to* S *such that* $\alpha_*\eta = \alpha$; *further,* α_* *is an isomorphism if and only if* α *is an epimorphism.*

Proof. Let k and l be any two elements of K, a any element of R. Then, since α is a homomorphism we have, as above

$$\alpha(k - l) = \alpha(k) - \alpha(l) = 0$$

and also

$$\alpha(ak) = \alpha(a)\,\alpha(k) = 0 = \alpha(k)\,\alpha(a) = \alpha(ka).$$

So $k - l$, ak and ka belong to K, which is therefore a two-sided ideal by Theorem 3.3.

We now define a mapping α_* from the residue class ring R/K to S as follows. Let C be any element of R/K and choose any element a of R lying in the residue class C; then set $\alpha_*(C) = \alpha(a)$. To show that $\alpha_*(C)$ so defined depends only on C and not on the choice of a, let a_1 be another element of C; then $a - a_1 \in K$ and hence $0 = \alpha(a - a_1) = \alpha(a) - \alpha(a_1)$, i.e. $\alpha(a) = \alpha(a_1)$, as required. It is clear from the definition of α_* that $\alpha_*\eta = \alpha$; since η is surjective it follows that α_* is the only mapping φ such that $\varphi\eta = \alpha$.

We prove next that α_* is a homomorphism. So let C and C' be any two elements of R/K and choose elements a, a' from C, C' respectively. Then

$$\alpha_*(C) + \alpha_*(C') = \alpha_*(\eta(a)) + \alpha_*(\eta(a')) = \alpha(a) + \alpha(a') = \alpha(a + a')$$
$$= \alpha_*(\eta(a + a')) = \alpha_*(C + C').$$

Similarly $\alpha_*(C)\alpha_*(C') = \alpha_*(CC')$. So α_* is a homomorphism, as asserted.

To show that α_* is a monomorphism, let C be any element of Ker α_*. If a is any element of C we have $0 = \alpha_*(C) = \alpha(a)$; so a belongs to Ker $\alpha = K$. It follows that C is the zero residue class, and hence, by Theorem 4.2, that α_* is a monomorphism.

It is clear from the definition of α_* that Im $\alpha_* =$ Im α. Thus α_* is an epimorphism, and hence an isomorphism, if and only if α is an epimorphism.

We call α_* the monomorphism *induced* by α.

Example 4. Let E be any set, A any ring and R the ring of mappings from E to A. Let x be an element of E and consider the mapping σ_x from R to A defined by setting $\sigma_x(\varphi) = \varphi(x)$ for every mapping φ from E to A. Then σ_x is a homomorphism from R to A; it is actually an epimorphism, since for every element a of A there exists at least one mapping φ from E to A such that $\varphi(x) = a$, namely the constant mapping φ_a. The kernel of σ_x is the ideal $I(x)$ described in Example 2 of §3. Thus we deduce from Theorem 4.3 that $R/I(x)$ is isomorphic to A.

The next homomorphism theorem is a simple application of Theorem 4.3.

THEOREM 4.4. (*Second homomorphism theorem*). *Let A and B be two-sided ideals of a ring R. Then there exists an isomorphism from $(A + B)/B$ onto $A/(A \cap B)$.*

Proof. We recall that $A + B$ is the subset of R consisting of all elements of the form $a + b$ with a in A and b in B. Clearly B is a two-sided ideal of $A + B$, and $A \cap B$ is a two-sided ideal of A.

Let η be the canonical epimorphism from A onto $A/(A \cap B)$, and consider the mapping α from $A + B$ to $A/(A \cap B)$ defined as follows: for each element x of $A + B$ write x in the form $x = a + b$, where $a \in A$ and $b \in B$, and set $\alpha(x) = \eta(a)$. We must check that the right hand member depends only on x and not on the particular representation of x. So suppose $x = a + b = a' + b'$ where $a, a' \in A$ and $b, b' \in B$. Then $a - a' = b' - b$; hence $a - a' \in A \cap B$ and so $\eta(a) = \eta(a')$, as we hoped.

To show that α is a homomorphism from $A + B$ to $A/(A \cap B)$ let $x_1 = a_1 + b_1, x_2 = a_2 + b_2$ be elements of $A + B$ (where $a_1, a_2 \in A$, $b_1, b_2 \in B$). Then we have

$$x_1 + x_2 = (a_1 + a_2) + (b_1 + b_2),$$

$$x_1 x_2 = a_1 a_2 + (a_1 b_2 + a_2 b_1 + b_1 b_2);$$

and, since A and B are two-sided ideals we have $a_1 + a_2 \in A$,

$a_1a_2 \in A$, $b_1 + b_2 \in B$ and $a_1b_2 + a_2b_1 + b_1b_2 \in B$. Hence we have

$$\alpha(x_1 + x_2) = \eta(a_1 + a_2) = \eta(a_1) + \eta(a_2) = \alpha(x_1) + \alpha(x_2)$$

and

$$\alpha(x_1x_2) = \eta(a_1a_2) = \eta(a_1)\,\eta(a_2) = \alpha(x_1)\,\alpha(x_2).$$

Since η is an epimorphism and $\alpha(a) = \eta(a)$ for every element a of A it follows that α is an epimorphism.

Now the element $x = a + b$ (where $a \in A$, $b \in B$) belongs to Ker α if and only if $\alpha(x) = \eta(a)$ is the zero element of $A/(A \cap B)$, i.e. if and only if $a \in A \cap B$. But since a is an element of A, we have $a \in A \cap B$ if and only if $a \in B$, and this happens if and only if $x = a + b$ belongs to B. Thus Ker $\alpha = B$.

The induced map α_* from $(A + B)/B$ to $A/(A \cap B)$ is then the required isomorphism.

The next main theorem on homomorphisms gives us a description of the ideals of any homomorphic image of a ring in terms of the ideals in the ring itself.

THEOREM 4.5. (*Third homomorphism theorem*). *Let α be an epimorphism from a ring R to a ring S with kernel K. Then there is a one-to-one correspondence between the set of ideals of R (left, right and two-sided) which include K and the set of all ideals of S. Further, if I is any two-sided ideal of R which includes K then R/I is isomorphic to $S/\alpha(I)$.*

Proof. Let $\mathscr{R}_K$ be the set of ideals of R which include K, $\mathscr{S}$ the set of all ideals of S. The mapping α from R to S gives rise to a mapping from $\mathscr{R}_K$ to the set of subsets of S; this is the mapping which assigns to each ideal I in $\mathscr{R}_K$ its image under the mapping α; we shall denote this mapping by α also. We have also a mapping β from $\mathscr{S}$ to the set of subsets of R defined by setting $\beta(J) = \alpha^{-1}(J)$ for each ideal J in $\mathscr{S}$. We shall show that α and β set up the required one-to-one correspondence between $\mathscr{R}_K$ and $\mathscr{S}$.

First we show that α actually maps $\mathscr{R}_K$ into $\mathscr{S}$. So let I be any ideal of R which includes K; let b_1, b_2 be elements of $\alpha(I)$, s any element of $S = \alpha(R)$. Then there are elements a_1, a_2 of I and r of R such that $b_1 = \alpha(a_1)$, $b_2 = \alpha(a_2)$, $s = \alpha(r)$. Hence we have

$$b_1 - b_2 = \alpha(a_1) - \alpha(a_2) = \alpha(a_1 - a_2),$$
$$sb_1 = \alpha(r)\,\alpha(a_1) = \alpha(ra_1), b_1s = \alpha(a_1)\,\alpha(r) = \alpha(a_1r),$$

from which it follows that $\alpha(I)$ is a left, right or two-sided ideal of S according as I is a left, right or two-sided ideal of R.

Next we show that β maps $\mathscr{S}$ into $\mathscr{R}_K$. First let J be any left ideal of S. Then certainly $\beta(J) = \alpha^{-1}(J)$ includes $\alpha^{-1}(0) = K$. Now let a_1, a_2 be elements of $\beta(J)$, r any element of R; then, by definition of β, $\alpha(a_1) \in J$, $\alpha(a_2) \in J$ and of course $\alpha(r) \in S$. We have $\alpha(a_1 - a_2) = \alpha(a_1) - \alpha(a_2) \in J$, so $a_1 - a_2 \in \beta(J)$; similarly $\alpha(ra_1) = \alpha(r)\,\alpha(a_1) \in J$ and so $ra_1 \in \beta(J)$. Thus $\beta(J)$ is a left ideal of R. Similar arguments show that if J is a right or two-sided ideal of S then $\beta(J)$ is a right or two-sided ideal of I which includes K.

Now we show that $\alpha\beta$ and $\beta\alpha$ are the identity mappings of $\mathscr{S}$ and $\mathscr{R}_K$ respectively.

If J is any ideal in $\mathscr{S}$ we certainly have $\alpha\beta(J) = \alpha(\alpha^{-1}(J)) \subseteq J$. Conversely, if b is any element of J then, since α is surjective, there is an element a of R such that $b = \alpha(a)$; but then $a \in \alpha^{-1}(b) \subseteq \alpha^{-1}(J)$ and so $b \in \alpha(\alpha^{-1}(J)) = \alpha\beta(J)$. Thus $J \subseteq \alpha\beta(J)$ and hence, as asserted, $\alpha\beta(J) = J$.

On the other hand, if I is any ideal in $\mathscr{R}_K$, we have $\beta\alpha(I) = \alpha^{-1}(\alpha(I)) \supseteq I$. If, conversely, $r \in \beta\alpha(I)$ it follows that $\alpha(r) \in \alpha(I)$; so there is an element a of I such that $\alpha(r) = \alpha(a)$, whence $\alpha(r - a) = 0$ and so $r - a \in K$. Since K is included in I it follows that $r - a \in I$ and hence that $r \in I$. Thus $\beta\alpha(I) \subseteq I$ and consequently $\beta\alpha(I) = I$.

It now follows that $\mathscr{R}_K$ and $\mathscr{S}$ are in one-to-one correspondence.

Finally let I be any two-sided ideal of R which includes K. Let η be the canonical epimorphism from S onto $S/\alpha(I)$, and consider the mapping $\varphi = \eta\alpha$ from R to $S/\alpha(I)$. It is easily verified that in fact φ is an epimorphism from R onto $S/\alpha(I)$. Clearly I is included in $\operatorname{Ker}\varphi$; but conversely, if $a \in \operatorname{Ker}\varphi$ we have $\alpha(a) \in \alpha(I)$ and as above we deduce that $a \in I$. So $\operatorname{Ker}\varphi = I$ and it follows from Theorem 4.3 that R/I is isomorphic to $S/\alpha(I)$, as asserted.

This completes the proof of the theorem.

Let R be a ring with identity element e. Consider the mapping φ from the ring of integers $\mathbf{Z}$ to R defined by setting $\varphi(z) = ze$ for every integer z. (For the definition of the integral multiples ze see §2.) Then it is easy to verify that φ is a homomorphism. The kernel of φ is an ideal of $\mathbf{Z}$ and hence, by Example 1 of §3, a principal ideal (n) where n is a positive integer or zero. This integer n is called the *characteristic* of the ring R.

Example 5. The rings $\mathbf{Z}, \mathbf{Q}, \mathbf{R}, \mathbf{C}, \mathbf{H}$ all have characteristic zero. For each positive integer m the residue class ring $\mathbf{Z}/(m)$ has characteristic m.

It follows that there are rings of all possible characteristics. For fields, division rings and integral domains, however, the possible characteristics are restricted, as the next theorem shows.

THEOREM 4.6. *Let R be a ring with identity. If R has no divisors of zero then the characteristic of R is either zero or a prime number.*

Proof. Suppose, to the contrary, that the characteristic n of R is neither zero nor a prime number. Then $n = n_1 n_2$ where $1 < n_1$, $n_2 < n$, and we have $0 = ne = (n_1 n_2)e = (n_1 e)(n_2 e)$. Hence, since R has no divisors of zero, either $n_1 e = 0$ or $n_2 e = 0$, i.e. either $n_1 \in \text{Ker } \varphi$ or $n_2 \in \text{Ker } \varphi$, where φ is the homomorphism described above. But this is a contradiction, since $\text{Ker } \varphi = (n)$ and n_1 and n_2 are not divisible by n. It follows that n is either zero or a prime number.

EXERCISES 1

1. Let φ be a multiplication operation on a set E; for each element a of E define the *left* and *right translations* by a to be the mappings λ_a and ρ_a from E to E defined by setting

$$\lambda_a(x) = ax \quad \text{and} \quad \rho_a(x) = xa$$

for all elements x of E. Prove

(a) φ is associative if and only if every left translation commutes with every right translation under composition of mappings in $\text{Map}(E, E)$;

(b) if λ_a and ρ_a are surjective there exists a neutral element with respect to φ and a is invertible.

2. Let A be a set with an addition operation and a multiplication operation satisfying conditions $A1$, $A3$, $A4$, $M1$, $M3$, AM, i.e. all the conditions for a ring with identity except the commutativity of addition. If a and b are any two elements of A expand $(a + b)(e + e)$ in two ways using AM (where e is the identity for multiplication) and deduce that A is actually a ring.

3. Prove that an integral domain with a finite number of elements is a field.

4. Show that the set of complex numbers of the form $a + b\omega$ where a and b are rational numbers and $\omega = \cos\frac{2}{3}\pi + i \sin\frac{2}{3}\pi$ is a subfield of $\mathbf{C}$.

5. Show that the set of matrices of the form $\begin{bmatrix} a & b \\ 0 & 0 \end{bmatrix}$ where a and b are real numbers is a subring of $M_2(\mathbf{R})$. Prove that it has no right neutral element for multiplication but infinitely many left neutral elements.

6. Let M be a subset of a ring R; let M^* be the set of elements of R which commute with every element of M. Prove that M^* is a subring of R. If S is a subring of R prove that S is included in S^{**} and deduce that $S^* = S^{***}$. The subring R^* is called the centre of R; prove that the centre of a division ring is a field. Find the centre of the ring $\mathbf{H}$ of real quaternions.

7. Show that the set of matrices of the form $\begin{bmatrix} p & q \\ 0 & s \end{bmatrix}$ where p, q, s are integers is a subring of $M_2(\mathbf{Z})$. Find all its ideals.

8. Let R be a commutative ring, A and B ideals of R. Prove that the set

$$(A:B) = \{x \in R : xb \in A \text{ for all } b \text{ in } B\}$$

is an ideal of R such that $(A:B)B \subseteq A \subseteq (A:B)$. Show also that $(A:B) = (A:(A + B))$ and that if C is a third ideal of R then $((A:B):C) = (A:BC)$.

9. Let ρ be an equivalence relation on a ring R which is compatible with the addition and multiplication operations in R. Prove that there exists a two-sided ideal I of R such that $x\rho y$ if and only if $x - y \in I$.

10. Let R be a ring, R' the two-sided ideal generated by the set of all elements of the form $xy - yx$ where $x, y \in R$. If I is any two-sided ideal of R prove that R/I is commutative if and only if I includes R'.

11. An element x of a ring R is said to be *idempotent* if $x^2 = x$. Find the idempotent elements of the residue class rings $\mathbf{Z}/(10)$, $\mathbf{Z}/(11)$, $\mathbf{Z}/(12)$.

12. An element x of a ring R is said to be *nilpotent* if there is a natural number n such that $x^n = 0$. Find the nilpotent elements of the residue class rings $\mathbf{Z}/(9)$, $\mathbf{Z}/(10)$, $\mathbf{Z}/(11)$, $\mathbf{Z}/(12)$.

13. Let R be a commutative ring. Prove that the set N of nilpotent elements of R is a (two-sided) ideal of R and that the residue class ring R/N has no non-zero nilpotent elements.

14. Show that the only ring homomorphisms from $\mathbf{Z}$ to $\mathbf{Z}$ are the zero homomorphism and the identity automorphism.

15. Let R be a ring; let S be the set of ordered pairs of the form (a, n) where $a \in R$ and $n \in \mathbf{Z}$, i.e. $S = R \times \mathbf{Z}$. Define addition and multiplication operations in S by setting

$$(a_1, n_1) + (a_2, n_2) = (a_1 + a_2, n_1 + n_2)$$

$$(a_1, n_1) \times (a_2, n_2) = (a_1 a_2 + n_1 a_2 + n_2 a_1, n_1 n_2)$$

for all $(a_1, n_1), (a_2, n_2)$ in S. Prove that under these laws of composition S is a ring with identity and that the mapping $\imath$ from R to S defined by setting $\imath(a) = (a, 0)$ for all elements a of R is a monomorphism from R onto a two-sided ideal of S.

CHAPTER 2
MODULES

§5. Modules, Submodules and Factor Modules

Let A and E be sets; by an *external law of composition on E with A as set of operators* we mean a mapping from the Cartesian product $A \times E$ to E. If μ is such an external law of composition and (a, x) is any element of $A \times E$, we shall usually denote the image $\mu(a, x)$ in E by ax or $a.x$, in which case μ is called *scalar multiplication on the left* of E by elements of A, or by xa or $x.a$ when μ is called *scalar multiplication on the right*.

Example 1. Let $A = \mathbf{N}$, the set of natural numbers, E a semigroup with identity under the internal law of composition $\wedge$. Then the mapping μ from $\mathbf{N} \times E$ to E given by $\mu(n, x) = \wedge^n x$ (as defined in §2) for every natural number n and every element x of E is an external law of composition on E with $\mathbf{N}$ as set of operators.

Example 2. Let $A = \mathbf{R}$, the set of real numbers, E the set of vectors in three-dimensional Euclidean space. For every non-zero real number a and every non-zero vector x in E let $\mu(a, x)$ be the vector whose magnitude is $|a| \, \|x\|$ (where $\|x\|$ is the magnitude of x), whose direction is the same as that of x and whose sense is the same as that of x or opposite to it according as a is positive or negative; if either a or x is zero, let $\mu(a, x)$ be the zero vector. Then the mapping μ from $\mathbf{R} \times E$ to E defined in this way is an external law of composition on E with $\mathbf{R}$ as set of operators.

Let R be a ring with identity element e; let V be a set with an internal law of composition φ which we write additively and an external law of composition μ with R as set of operators which we write as scalar multiplication on the left. Then we say that V is a *left R-module* under φ and μ or that (V, φ, μ) is a left R-module if the following conditions are satisfied:

A. V is an abelian group under the addition operation φ;
LM. For all elements a, b of R and x, y of V,

$$(1)\ (a + b)\, x = ax + bx,$$

(2) $a(x + y) = ax + ay$,
(3) $(ab) x = a(bx)$,
(4) $ex = x$.

If V has an addition operation φ and a right scalar multiplication μ by elements of R then we say that V is a *right R-module* under φ and μ if:

A. V is an abelian group under φ;
RM. For all elements a, b of R and x, y of V,

(1) $x(a + b) = xa + xb$,
(2) $(x + y) a = xa + ya$,
(3) $x(ab) = (xa) b$,
(4) $xe = x$.

If R were an arbitrary ring (i.e. not necessarily with an identity element) we might have defined left R-modules by postulating all the above requirements except for $LM(4)$; in the case where R had an identity left R-modules satisfying $LM(4)$ would be called unitary modules. Similar remarks apply of course to right R-modules with reference to $RM(4)$. We have chosen to assume from now on that *all rings have identity elements and all modules are unitary*; so we have built $LM(4)$ and $RM(4)$ into our basic definition of modules.

If R is a division ring (or, in particular, a field) left R-modules are usually referred to as *left vector spaces* over R and right R-modules are called *right vector spaces* over R.

Example 3. Let A be any additive abelian group (i.e. an abelian group with its law of composition written additively). Let μ be the mapping from $\mathbf{Z} \times A$ to A given by setting $\mu(z, a) = za$ (as defined in §2) for every integer z and every element a of A. Then A is a left $\mathbf{Z}$-module under the addition in A and the mapping μ.

Example 4. Let E be the set of vectors in three-dimensional Euclidean space. Then E is a left vector space over the real number field $\mathbf{R}$ under ordinary vector addition and the scalar multiplication μ defined in Example 2.

Example 5. Let A be any additive abelian group, and let $R = \text{End } A$ be the ring of endomorphisms of A as defined in §2, Example 6. Define the mapping μ from $R \times A$ to A by setting $\mu(\alpha, a) = \alpha(a)$ for every endomorphism α and every element a of A. Then it is an easy matter to verify that A is a left R-module under its addition operation and the external law μ.

Example 6. Let R be any ring with identity. Consider the mappings λ and ρ from $R \times R$ to R defined by setting $\lambda(a, x) = ax$ and $\rho(a, x) = xa$ for all elements a and x of R. Then R is a left R-module under its addition operation and the 'external law' λ, and a right R-module under its addition operation and ρ. When we consider R as a left or right R-module in this way we shall denote it by R_l or R_r respectively.

During the elementary development of the theory of modules we shall give our definitions and results for left modules, leaving it to the reader to make the obvious modifications necessary to deal with right modules.

When we are working with a left R-module V two zero elements will occur in our discussions—the zero element of the ring R and the zero element of the additive group V. We shall usually denote both these elements simply by 0; but, if it is necessary to emphasise which we are dealing with we may denote them by 0_R and 0_V respectively.

If x is any element of V we claim that $0_R x = 0_V$. To see this we remark that $0_R + 0_R = 0_R$ and hence $(0_R + 0_R) x = 0_R x$; by $LM(1)$ we deduce that $0_R x + 0_R x = 0_R x$. Adding the negative $-(0_R x)$ to both sides and using the associativity of the addition operation in V, we obtain $0_R x = 0_V$. A similar argument shows that if a is any element of R then $a0_V = 0_V$. Now let a be any element of R, x any element of V; we claim that $(-a) x = -(ax)$. To show this we have only to remark that $ax + (-a) x = (a + (-a)) x = 0_R x = 0_V$; hence $(-a) x$ is an inverse for ax with respect to the addition in V. But so is $-(ax)$; and since inverses in a group are unique, we have $(-a) x = -(ax)$. Similarly $a(-x) = -(ax)$.

In §1 we described how to form the sum $\sum_{i \in I} x_i$ of a *finite* family $(x_i)_{i \in I}$ of elements of a left R-module V. Now, suppose that the index set I is quite arbitrary, i.e. not necessarily finite, and let $(x_i)_{i \in I}$ be a family of elements of V. Let I^* be the subset of I consisting of indices i for which x_i is non-zero. Then if I^* is finite but non-empty, $\sum_{i \in I} x_i$ is defined to mean $\sum_{i \in I^*} x_i$ (as described in §1); if I^* is empty, $\sum_{i \in I} x_i$ is defined to be the zero element of V; and if I^* is infinite $\sum_{i \in I} x_i$ is left undefined.

Let V be a left R-module, W a subset of V. We say that W is *closed* or *stable* under the scalar multiplication μ of V by R if for every element a of R and every element w of W we have $\mu(a, w) = aw \in W$. In this situation we say that μ *induces* an external law of composition on W. If, further, W is closed under the addition operation in V (which therefore induces an addition on W) we may sensibly ask whether W forms a left R-module under these induced laws of composition. If it does, we say that W is a *submodule* of V.

Example 7. Let A be any additive abelian group considered as a left **Z**-module according to the description in Example 3. Then every subgroup of A is a submodule of A.

Example 8. As in Example 6 let R_l be the ring R itself considered as a left R-module. Then it is easily verified that the submodules of R_l are just the left ideals of the ring R considered as left R-modules.

The situation here is reminiscent of that in §3 where we considered subsets of a ring which were closed under the laws of composition of the ring. We saw that certain additional conditions, over and above the closure, were necessary and sufficient for such a subset to be a ring. In the case of modules, however, we shall show that the closure conditions are already themselves necessary and sufficient for a subset to be a submodule.

THEOREM 5.1. *A non-empty subset W of a left R-module V is a submodule if and only if it is closed under the addition in V and the scalar multiplication of V by R.*

Proof. (1) If W is a submodule of V it is certainly closed under the addition and scalar multiplication.

(2) Conversely, suppose W is closed under these laws of composition. The conditions $LM(1)$–(4) are certainly satisfied by the induced scalar multiplication on W since they are satisfied by the original scalar multiplication in V.

All that now remains to be shown is that W is an abelian group under the induced addition. This induced addition is certainly associative and commutative, since the original addition in V is associative and commutative. Let w be any element of W; since W is closed under the scalar multiplication of V by R, $0_R w = 0_V$ belongs to W, which therefore has a neutral element for addition.

Again let w be any element of W; using once more the fact that W is closed under the scalar multiplication of V by R we deduce that $(-e)w = -(ew) = -w$ belongs to W. Thus W is indeed an abelian group under the induced addition.

This completes the proof.

COROLLARY. *The intersection of any non-empty family of submodules of a left R-module V is also a submodule of V.*

The module V itself is clearly a submodule of V; the submodules of V distinct from V are called *proper submodules* of V. If V does not consist of the zero element 0_V alone then $\{0_V\}$ is a proper submodule of V, which we call the *zero submodule*. A left R-module V is said to be *simple* or *irreducible* if (1) it does not consist of the zero element alone and (2) its only submodules are V itself and the zero submodule.

Let X be any subset of a left R-module V. Then the set of all submodules of V which include X is certainly non-empty, since the module V itself belongs to this set. According to the Corollary of Theorem 5.1. the intersection of this set is a submodule of V, and it certainly includes X; this is clearly the smallest submodule of V which includes X. We denote it by RX (or by Rx if X consists of a single element x of V) and call it the submodule of V *generated by* X (or by x). If X is a subset of V such that $RX = V$ we call X a *generating system* for V; if V has a finite generating system we say that V is a *finitely generated* module.

In order to describe the elements of RX it is convenient to introduce the following terminology: an element y of V is said to be expressed as a *linear combination* of elements of X with coefficients in R if there exists a family $(a_x)_{x \in X}$ of elements of R such that only finitely many of the elements a_x are non-zero and $y = \sum_{x \in X} a_x x$.

We shall have to mention very frequently families of elements of R only finitely many of which are non-zero: we call such families *quasi-finite*. Thus a family $(a_i)_{i \in I}$ of elements of R is quasi-finite if and only if the subset I^* of I consisting of those indices i for which a_i is non-zero is a finite set.

THEOREM 5.2. *Let X be a subset of a left R-module V. Then the submodule RX of V generated by X consists of all the elements of V which can be expressed as linear combinations of elements of X with coefficients in R.*

Proof. Let W be the subset of V consisting of all these linear combinations. Then certainly W is included in every submodule of V which includes X and hence $W \subseteq RX$.

But it follows at once from Theorem 5.1 that W is itself a submodule of V, and it clearly includes X. So $W \supseteq RX$.

Hence $W = RX$ as asserted.

By specialising to the case where X consists of a single element x of V we obtain the following consequence.

COROLLARY. *Let x be an element of a left R-module V. Then the submodule Rx of V generated by x consists of all the elements of V of the form ax where a is an element of R.*

Let $(W_i)_{i \in I}$ be a family of submodules of a left R-module V. We may form the union $\cup_{i \in I} W_i$ of this family; this union is of course a subset of V, but it is not in general a submodule. Nevertheless $\cup_{i \in I} W_i$ generates a submodule of V, which we denote by $\sum_{i \in I} W_i$ and call the *sum* of the family $(W_i)_{i \in I}$. In the same way if W_1 and W_2 are submodules of V the submodule generated by $W_1 \cup W_2$ is denoted by $W_1 + W_2$. By using an argument similar to that in the proof of Theorem 5.2 we obtain the following description of the elements of $\sum_{i \in I} W_i$.

THEOREM 5.3. *Let $(W_i)_{i \in I}$ be a family of submodules of a left R-module V. Then the sum $\sum_{i \in I} W_i$ of this family consists of all elements of V of the form $\sum_{i \in I} w_i$ where $w_i \in W_i$ for each index i in I and only finitely many of the elements w_i are non-zero.*

A finite subset X of a left R-module V is said to be *linearly dependent* if there exists a family $(a_x)_{x \in X}$ of elements of R, *not all zero*, such that $\sum_{x \in X} a_x x = 0$. An arbitrary subset is said to be linearly dependent if it has a finite subset which is linearly dependent; that is to say, X is linearly dependent if there is a linear combination of elements of X with non-zero coefficients in R which is equal to the zero element of W. The subset X is said to be *linearly independent* or *free* if it is not linearly dependent; thus X is free if and only if the only linear combination of elements of X with coefficients in R which can be equal to zero is that in which all the coefficients are zero. We make the convention that the empty set is free. It is clear that a subset of V is free if and only if all its finite subsets are free.

If $RX = V$ and X is linearly independent, i.e. if X is a free generating system for V, we say that X is a *basis* for V. The next theorem gives a useful criterion for a subset of V to be a basis.

THEOREM 5.4. *A subset X of a left R-module V is a basis for V if and only if every element of V can be expressed uniquely as a linear combination of elements of X with coefficients in R.*

Proof. (1) Suppose X is a basis for V.

Since $RX = V$ it follows from Theorem 5.2 that every element of V is expressible as a linear combination of elements of X.

To show that the expression is unique, let y be any element of V and suppose there are quasi-finite families $(a_x)_{x \in X}$ and $(b_x)_{x \in X}$ of elements of R such that $y = \sum_{x \in X} a_x x = \sum_{x \in X} b_x x$. Then we have $\sum_{x \in X} (a_x - b_x) x = 0$ and hence, since X is linearly independent, $a_x - b_x = 0$, i.e. $a_x = b_x$ for all elements x of X as required.

(2) Conversely, suppose every element of V can be expressed uniquely as a linear combination of elements of X with coefficients in R.

It follows at once from Theorem 5.2 that $RX = V$.

To show that X is linearly independent, let $(a_x)_{x \in X}$ be a quasi-finite family of elements of R such that $\sum_{x \in X} a_x x = 0$. It is clear, however, that if $b_x = 0$ for every element x of X we have $\sum_{x \in X} b_x x = 0$ also. It now follows from the uniqueness assumption that $a_x = b_x = 0$ for every element x of X; so X is linearly independent.

Example 9. Let $V = \mathbf{Z}_l$, i.e., the ring of integers considered as a left **Z**-module. Then the set $X = \{1\}$ is a basis for V. First of all X generates V; for if $z \in \mathbf{Z}$ we have $z = z1 \in \mathbf{Z}X$. Secondly, X is free; for if we have $z1 = 0$ it follows that $z = 0$. We may show in the same way that $\{-1\}$ is a basis for V.

Example 10. Let V be a finite abelian group, considered as a **Z**-module in the way we described in Example 3. Then V has no basis. For if n is the order of V, then, according to a well-known result of elementary group theory, we have $nx = 0$ for every element x of V. Thus V cannot have any non-empty free subsets.

This last example shows that we cannot assert in general that every left R-module has a basis. We shall prove, however, that if R is a division ring (or a field) then every R-module (i.e. every vector space over R) has a basis.

THEOREM 5.5. *Let V be a left R-module over a division ring R. If X is a free subset of V there exists a basis for V including X.*

Proof. Let F be the set of linearly independent subsets of V which include X. Then the inclusion relation between subsets of V is an order relation on F. We claim that F is inductively ordered by this relation.

So let F_0 be a totally ordered subset of F; we must show that F_0 has a least upper bound in F. If X_0 is the union of the sets in F_0, then X_0 is clearly the least upper bound of F_0 in the set of all subsets of V which include X; we claim that in fact X_0 belongs to F.

Suppose, to the contrary, that X_0 is not free. Then there is a finite subset of X_0, say $\{x_1, \ldots, x_k\}$, which is linearly dependent. Since X_0 is the union of the sets in F_0, it follows that there are sets $X_{i_1}, \ldots, X_{i_k}$ in F_0 such that $x_j \in X_{i_j}$ ($j = 1, \ldots, k$). Now F_0 is totally ordered; so one of the sets $X_{i_1}, \ldots, X_{i_k}$ in F_0, say X_{i_1}, includes all the others. But then the free subset X_{i_1} includes the linearly dependent set $\{x_1, \ldots, x_k\}$; and this is a contradiction. Hence X_0 is free.

Applying Zorn's Lemma we deduce that F has a maximal element, i.e. that there is at least one maximal linearly independent subset of V which includes X; choose one such subset and call it B. The argument so far has made no use of the fact that R is a division ring; but this fact is crucial for our next step in which we show that B is a basis for V.

Since B is certainly free, all that remains is to show that B is a generating system for V, in other words that $RB = V$. Suppose then that $RB \neq V$ and let y be any element of V which does not belong to RB. We claim that $B \cup \{y\}$ is linearly independent; so let $ay + \sum_{x \in B} a_x x = 0$, where $a \in R$ and $(a_x)_{x \in B}$ is a quasi-finite family of elements of R. If $a \neq 0$ then, since R is a division ring, a has a multiplicative inverse and we deduce that $y = \sum_{x \in B} (-a^{-1} a_x) x \in RB$, contradicting our assumption that $y \notin RB$. Hence $a = 0$ and so $\sum_{x \in B} a_x x = 0$, from which it follows that all the coefficients a_x are zero (since B is free). Thus $B \cup \{y\}$ is linearly independent. But this contradicts the maximal property of B. Hence $RB = V$ and so B is a basis for V, as required.

COROLLARY. *If R is a division ring then every left R-module has a basis.*

Proof. We have only to apply the theorem in the case where $X = \phi$, remembering that ϕ is defined to be a free subset.

Let V be a left R-module, W a submodule of V. We define a relation on V by setting $x \equiv y$ (mod. W) if and only if $x - y \in W$, and verify that this is an equivalence relation. The addition operation in V is compatible with this relation, for if $x \equiv x'$ (mod. W) and $y \equiv y'$ (mod. W) we have $x - x' \in W$ and $y - y' \in W$, whence $(x + y) - (x' + y') = (x - x') + (y - y') \in W$ (since the submodule W is closed under the addition in V); thus $x + y \equiv x' + y'$ (mod. W). The quotient set of V with respect to this equivalence relation is denoted by V/W and the equivalence classes are called the *cosets* of V modulo the submodule W; the canonical surjection from V onto V/W is denoted by η.

According to §1 the addition operation in V induces an addition operation in V/W as follows: if C_1, C_2 are cosets modulo W and x_1, x_2 are elements of V in C_1, C_2 respectively, then $C_1 + C_2 = \eta(x_1 + x_2)$; we recall that since the addition in V is compatible with our equivalence relation the coset $C_1 + C_2$ depends only on C_1 and C_2, not on the choices of x_1 and x_2. It is easily verified that V/W forms an abelian group under the induced addition: the associativity and commutativity follow from the corresponding conditions in V; $\eta(0_V) = W$ is a zero element for V/W, and if $C = \eta(x)$ is any coset of V modulo W, $\eta(-x)$ is an additive inverse for C.

We now define a left scalar multiplication of V/W by elements of R. Let C be any coset of V modulo W; choose any element x of C. Then for each element a of R define aC to be the coset $\eta(ax)$. We claim that this coset depends only on a and C, not on the choice of x. For if x' is another element of V in C, we have $x \equiv x'$ (mod. W), i.e. $x - x' \in W$; hence, since W is closed under the scalar multiplication in V, $ax - ax' = a(x - x') \in W$. So $ax \equiv ax'$ (mod. W) and $\eta(ax) = \eta(ax')$ as required. A routine verification shows that V/W is a left R-module under the induced addition and the scalar multiplication we have just defined. We call V/W the *factor module* or *quotient module* of V modulo W.

Example 11. Let A be an additive abelian group considered as a **Z**-module according to Example 3. If B is any subgroup of A (submodule of A as **Z**-module) the factor module is simply the factor group of A modulo B regarded as a **Z**-module in the usual way.

Example 12. If R is any ring, A any left ideal of R, then A is a submodule of the left R-module R_l. Thus we may form the factor module R_l/A, which is a left R-module. We must distinguish this construction carefully from that of the residue class ring R/I described in §3—the latter can be carried out only if I is a two-sided ideal of R.

§6. Homomorphisms of Modules

Let R be a ring (with identity, as usual), and let V and V' be (unitary) left R-modules. Then a mapping α from V to V' is called an *R-module homomorphism* (or an R-homomorphism or simply a homomorphism), if for all elements x, y of V and all elements a of R we have

$$\alpha(x + y) = \alpha(x) + \alpha(y) \qquad \text{and} \qquad \alpha(ax) = a\alpha(x).$$

The other terms monomorphism, epimorphism,... defined in §1 are all applied in the obvious way to special types of module homomorphisms. As usual the composition of two homomorphisms is a homomorphism and the inverse of an isomorphism is an isomorphism.

Example 1. Let V and V' be any two left R-modules. The mapping ζ from V to V' defined by setting $\zeta(x) = 0$ for all elements x of V is a module homomorphism from V to V', which we call the *zero homomorphism.*

Example 2. Let A and A' be abelian groups, α a group homomorphism from A to A'. Then α is also a **Z**-module homomorphism from A to A' when these groups are considered as **Z**-modules in the usual way.

Example 3. Let V be any left R-module, W a submodule of V, ι the canonical injection from W to V. Then ι is a monomorphism from W to V, which we call the *inclusion monomorphism.* In particular, if $W = V$ the identity mapping I_V is an automorphism of V.

Example 4. Let V be any left R-module, W a submodule of V. Then the addition operation and scalar multiplication in the factor module V/W are defined in such a way that the canonical surjection η from V onto V/W is a module homomorphism. From now on we shall call η the *canonical epimorphism* from V to V/W.

The basic results about module homomorphisms are closely analogous to those for ring homomorphisms and we shall omit most of the proofs; the interested reader will find it easy to construct them for himself if he follows, with minor modifications, the models given in §4.

THEOREM 6.1. *Let V and V' be left R-modules, α a homomorphism from V to V'. Then* $\text{Im } \alpha = \alpha(V)$ *is a submodule of V'. If 0_V is the zero element of V then $\alpha(0_V) = 0_{V'}$, the zero element of V'; if x is any element of V then $\alpha(-x)$ is the additive inverse of $\alpha(x)$ in V'.*

The *kernel* of a module homomorphism α from V to V' is defined as for ring homomorphisms: it consists of the elements of V which are mapped by α onto the zero element of V'; we denote the kernel of α by $\text{Ker } \alpha$. Since $0_V \in \text{Ker } \alpha$ for every homomorphism α, the kernel of a homomorphism is always non-empty. We have the familiar criterion for a homomorphism to be injective.

THEOREM 6.2. *Let V and V' be left R-modules, α a homomorphism from V to V'. Then α is a monomorphism if and only if $\text{Ker } \alpha = \{0_V\}$.*

Next we have the fundamental theorem on module homomorphisms.

THEOREM 6.3. (*First homomorphism theorem*). *Let V and V' be left R-modules, α a homomorphism from V to V' with kernel V_0. Then V_0 is a submodule of V and if η is the canonical epimorphism from V onto V/V_0 there exists a unique monomorphism α_* from V/V_0 to V' such that $\alpha_* \eta = \alpha$; α_* is an isomorphism if and only if α is an epimorphism.*

Proof. We define a mapping α_* from V/V_0 to V' in the familiar way: for each coset C of V modulo V_0 choose an element x from C and set $\alpha_*(C) = \alpha(x)$. The right hand member $\alpha(x)$ depends only on C, not on the choice of x; for if $x' \in C$, then $x \equiv x'$ (mod. V_0), whence $x - x' \in V_0$, and so eventually $\alpha(x) = \alpha(x')$. If x is any element of V then since $x \in \eta(x)$ we certainly have $\alpha_*(\eta(x)) = \alpha(x)$; thus $\alpha_* \eta = \alpha$ as required.

We have now to show that α_* is a module homomorphism from

V/V_0 to V'. So let C_1, C_2 be any two cosets of V modulo V_0, x_1, x_2 elements of C_1, C_2 respectively, and a any element of R. Then

$$\alpha_*(C_1 + C_2) = \alpha_*(\eta(x_1) + \eta(x_2)) = \alpha_*(\eta(x_1 + x_2)) = \alpha(x_1 + x_2)$$

while

$$\alpha_*(C_1) + \alpha_*(C_2) = \alpha_*(\eta(x_1)) + \alpha_*(\eta(x_2)) = \alpha(x_1) + \alpha(x_2) = \alpha(x_1 + x_2),$$

and

$$\alpha_*(aC_1) = \alpha_*(a\eta(x_1)) = \alpha_*(\eta(ax_1)) = \alpha(ax_1)$$
$$= a\alpha(x_1) = a\alpha_*(\eta(x_1)) = a\alpha_*(C_1).$$

Now the coset $C = \eta(x)$ belongs to the kernel of α_* if and only if $\alpha_*(C) = \alpha_*(\eta(x)) = \alpha(x) = 0$, i.e. if and only if $x \in \operatorname{Ker} \alpha$ and so $\eta(x)$ is the zero coset modulo V_0. Hence, by Theorem 6.2, α_* is a monomorphism.

The uniqueness of α_* and the fact that α_* is surjective if and only if α is surjective are established by the arguments used in the proof of Theorem 4.3.

As in §4, we deduce the other two basic homomorphism theorems.

THEOREM 6.4. (*Second homomorphism theorem*). *Let W_1 and W_2 be submodules of a left R-module V. Then there exists an isomorphism from $(W_1 + W_2)/W_2$ onto $W_1/(W_1 \cap W_2)$.*

THEOREM 6.5. (*Third homomorphism theorem*). *Let V and V' be left R-modules, α an epimorphism from V onto V' with kernel V_0. Then there is a one-to-one correspondence between the set of submodules of V which include V_0 and the set of all submodules of V'. Further, if V_1 is any submodule of V which includes V_0, then V/V_1 is isomorphic to $V'/\alpha(V_1)$.*

If V and V' are left R-modules and α is a homomorphism from V to V' the factor modules $V/\operatorname{Ker} \alpha$ and $V'/\operatorname{Im} \alpha$ are called the *coimage* and *cokernel* of the homomorphism α and are denoted by Coim α and Coker α respectively. Clearly α is an epimorphism if and only if Coker $\alpha = \{0_{V'}\}$ (cf. Theorem 6.2). Further, according to the First Homomorphism Theorem there exists an isomorphism α_* from Coim α to Im α.

Let A, B, C be three left R-modules and let α, β be homomorphisms from A to B and B to C respectively. Then the diagram

$$A \xrightarrow{\alpha} B \xrightarrow{\beta} C$$

is said to be *exact* if Im $\alpha = $ Ker β. If $n \geqslant 2$ the diagram

$$A_0 \xrightarrow{\alpha_0} A_1 \xrightarrow{\alpha_1} A_2 \xrightarrow{\alpha_2} A_3 \xrightarrow{\alpha_3} \ldots \xrightarrow{\alpha_{n-1}} A_n \qquad [6.1]$$

of left R-modules and homomorphisms is said to be *exact at* A_i $(i = 1, 2, \ldots, n - 1)$ if the diagram

$$A_{i-1} \xrightarrow{\alpha_{i-1}} A_i \xrightarrow{\alpha_i} A_{i+1} \qquad [6.2]$$

is exact. The diagram $[6.1]$ is said to be an *exact sequence* if it is exact at each module A_i $(i = 1, 2, \ldots, n - 1)$. In the same way an infinite diagram

$$A_0 \xrightarrow{\alpha_0} A_1 \xrightarrow{\alpha_1} A_2 \xrightarrow{\alpha_2} A_3 \xrightarrow{\alpha_3} \ldots \xrightarrow{\alpha_{n-1}} A_n \xrightarrow{\alpha_n} \ldots$$

of left R-modules and homomorphisms is said to be exact at A_i $(i \geqslant 1)$ if the diagram $[6.2]$ is exact, and to be an exact sequence if it is exact at every module A_i $(i \geqslant 1)$. We make similar definitions for infinite diagrams such as

$$\ldots \xrightarrow{\alpha_{-n-1}} A_{-n} \xrightarrow{\alpha_{-n}} \ldots \xrightarrow{\alpha_{-4}} A_{-3} \xrightarrow{\alpha_{-3}} A_{-2} \xrightarrow{\alpha_{-2}} A_{-1} \xrightarrow{\alpha_{-1}} A_0$$

and

$$\ldots \xrightarrow{\alpha_{-n-1}} A_{-n} \xrightarrow{\alpha_{-n}} \ldots \xrightarrow{\alpha_{-2}} A_{-1} \xrightarrow{\alpha_{-1}} A_0 \xrightarrow{\alpha_0} A_1 \xrightarrow{\alpha_1} \ldots \xrightarrow{\alpha_{n-1}} A_n \xrightarrow{\alpha_n} \ldots$$

Before proceeding to give examples of exact sequences we introduce some conventions of notation which we shall use constantly. First of all, we agree that when a left R-module V_0 consists of a zero element 0 alone then we shall write $V_0 = 0$ instead of the more strictly correct $\{0\}$. Next, if α is a homomorphism from such a zero module to any left R-module V we must have $\alpha(0) = 0_V$; that is to say, there is only one possible homomorphism from 0 to V. So we agree that whenever a homomorphism from 0 to V occurs in any of our diagrams of modules and homomorphisms we shall represent it simply by $0 \to V$ without giving a name to the homomorphism. Finally, if β is a homomorphism from any module V to 0, it must be the zero homomorphism, given by $\beta(x) = 0$ for every element x in V. We shall represent this by $V \to 0$, again without naming the homomorphism.

Example 5. Let V' and V be left R-modules, α a homomorphism from V' to V. Consider the diagram

$$0 \to V' \xrightarrow{\alpha} V. \qquad [6.3]$$

If this is an exact sequence the kernel of α coincides with the image of the first homomorphism, which consists of the zero element of V' alone. Thus, by Theorem 6.2, α is a monomorphism. Conversely, if α is a monomorphism, [6.3] is an exact sequence.

Example 6. Let V and V'' be left R-modules, β a homomorphism from V to V''. Consider the diagram

$$V \xrightarrow{\beta} V'' \to 0. \qquad [6.4]$$

If this sequence is exact the image of β coincides with the kernel of the second homomorphism, which is of course V'' itself. Thus Im $\beta = V''$, i.e. β is an epimorphism. Conversely, if β is an epimorphism, [6.4] is an exact sequence.

A diagram of the form

$$0 \to V' \xrightarrow{\alpha} V \xrightarrow{\beta} V'' \to 0 \qquad [6.5]$$

which is exact at V', V and V'' is called a *short exact sequence*. According to Examples 5 and 6 the homomorphisms α and β of [6.5] are respectively a monomorphism and an epimorphism.

Example 7. Let W be a submodule of a left R-module V, ι the inclusion monomorphism from W to V and η the canonical epimorphism from V onto V/W. Then the diagram

$$0 \to W \xrightarrow{\iota} V \xrightarrow{\eta} V/W \to 0$$

is a short exact sequence. It is easy to show that all short exact sequences are 'essentially' of this form; more formally, if we are given a short exact sequence [6.5], we can show that V' is isomorphic to a submodule W of V and V'' is isomorphic to the corresponding factor module V/W.

Example 8. Let V_1 and V_2 be left R-modules, α a homomorphism from V_1 to V_2. Let ι be the inclusion monomorphism from Ker α to V_1 and η the canonical epimorphism from V_2 onto Coker α. Then the sequence

$$0 \to \text{Ker } \alpha \xrightarrow{\iota} V_1 \xrightarrow{\alpha} V_2 \xrightarrow{\eta} \text{Coker } \alpha \to 0$$

is exact.

Example 9. As an illustration of the technique known as ‘diagram-chasing’ we prove half of the so-called *Five Lemma*. Consider the diagram

$$
\begin{array}{ccccccccc}
V_{-2} & \xrightarrow{\alpha_{-2}} & V_{-1} & \xrightarrow{\alpha_{-1}} & V_0 & \xrightarrow{\alpha_0} & V_1 & \xrightarrow{\alpha_1} & V_2 \\
\downarrow{\varphi_{-2}} & & \downarrow{\varphi_{-1}} & & \downarrow{\varphi_0} & & \downarrow{\varphi_1} & & \downarrow{\varphi_2} \\
W_{-2} & \xrightarrow{\beta_{-2}} & W_{-1} & \xrightarrow{\beta_{-1}} & W_0 & \xrightarrow{\beta_0} & W_1 & \xrightarrow{\beta_1} & W_2
\end{array}
$$

in which both rows are exact sequences of left R-modules and homomorphisms, and all four squares are commutative (i.e. $\varphi_{i+1}\alpha_i = \beta_i\varphi_i$ for $i = -2, -1, 0, 1$). Suppose we are given that φ_2 is a monomorphism and φ_{-1} and φ_1 are epimorphisms; we shall show that φ_0 is an epimorphism.

So let w_0 be any element of W_0. Since φ_1 is an epimorphism there is an element v_1 of V_1 such that $\varphi_1(v_1) = \beta_0(w_0)$. Then we have $\beta_1\varphi_1(v_1) = \beta_1\beta_0(w_0) = 0$, since the lower row is exact; hence $\varphi_2\alpha_1(v_1) = 0$, by the commutativity in the right-hand square. Now φ_2 is a monomorphism, so $\alpha_1(v_1) = 0$; thus $v_1 \in \operatorname{Ker}\alpha_1 = \operatorname{Im}\alpha_0$, and there is an element v_0 of V_0 such that $v_1 = \alpha_0(v_0)$. Then $\beta_0(w_0) = \varphi_1(v_1) = \varphi_1\alpha_0(v_0) = \beta_0\varphi_0(v_0)$, from which we deduce that $w_0 - \varphi_0(v_0) \in \operatorname{Ker}\beta_0 = \operatorname{Im}\beta_{-1}$. Hence there is an element w_{-1} of W_{-1} such that $w_0 - \varphi_0(v_0) = \beta_{-1}(w_{-1})$. Since φ_{-1} is an epimorphism, there is an element v_{-1} of V_{-1} such that $w_{-1} = \varphi_{-1}(v_{-1})$. Consequently we have $w_0 = \varphi_0(v_0) + \beta_{-1}(w_{-1}) = \varphi_0(v_0) + \beta_{-1}\varphi_{-1}(v_{-1}) = \varphi_0(v_0) + \varphi_0\alpha_{-1}(v_{-1}) \in \operatorname{Im}\varphi_0$. That is to say φ_0 is an epimorphism.

The reader is invited to try his hand at proving the other half of the Five Lemma: if φ_{-2} is an epimorphism and φ_{-1} and φ_1 are monomorphisms then φ_0 is a monomorphism.

§7. Groups of Homomorphisms

Let R be a ring with identity, V and W two left R-modules. We denote the set of R-homomorphisms from V to W by $\operatorname{Hom}_R(V, W)$ or simply by $\operatorname{Hom}(V, W)$ when the ring R is clear from the context. We define an operation of addition in $\operatorname{Hom}(V, W)$ as follows: for every pair α, β of homomorphisms from V to W consider the mapping

$\alpha + \beta$ from V to W defined by setting

$$(\alpha + \beta)(x) = \alpha(x) + \beta(x)$$

for each element x of V. It is easy to verify that this mapping $\alpha + \beta$ is actually a homomorphism from V to W; so we have in fact defined an internal law of composition in Hom (V, W). Routine calculation shows that this law of composition is associative and commutative; the zero homomorphism ζ from V to W is clearly a neutral element; and for each homomorphism α from V to W the mapping $-\alpha$ from V to W given by

$$(-\alpha)(x) = -\alpha(x)$$

for each element x of V is a homomorphism and $\alpha + (-\alpha) = \zeta$. Hence, under the addition operation we have defined, Hom (V, W) is an abelian group.

Suppose that R is a commutative ring and that V and W are left R-modules. In this case we define a scalar multiplication on the left of Hom (V, W) by elements of R as follows: for each element a of R and each homomorphism α from V to W, let $a\alpha$ be the mapping from V to W defined by setting

$$(a\alpha)(x) = a\alpha(x)$$

for each element x of V. We contend that this mapping $a\alpha$ is a homomorphism from V to W. To see this, let x and y be any two elements of V, b any element of R; then

$$(a\alpha)(x + y) = a\alpha(x + y) = a(\alpha(x) + \alpha(y)) =$$
$$a\alpha(x) + a\alpha(y) = (a\alpha)(x) + (a\alpha)(y)$$

and

$$(a\alpha)(bx) = a\alpha(bx) = a(b\alpha(x)) = b(a\alpha(x)) = b((a\alpha)(x)).$$

It is easily verified that with the addition defined in the preceding paragraph and this scalar multiplication Hom (V, W) becomes a left R-module. We observe that in proving that $a\alpha$ is an R-homomorphism we made essential use of the commutativity of R. Thus in general (i.e. when R is not necessarily commutative) Hom (V, W) is only an additive abelian group, not a left R-module.

There are two special cases, however, even when R is not commutative, in which additional structure can be imposed on the abelian group Hom (V, W): these are the cases in which $V = R_l$

(i.e. the ring R itself, considered as left R-module) and $W = R_l$. We consider these two cases in turn.

If a is any element of R, α any homomorphism from R_l to W, we consider the mapping $a\alpha$ from R to W defined by setting

$$(a\alpha)(r) = \alpha(ra) \qquad [7.1]$$

for every element r of R. Then $a\alpha$ is also a homomorphism from R_l to W; for if r_1, r_2, r, a_1 are any elements of R we have

$$(a\alpha)(r_1 + r_2) = \alpha((r_1 + r_2)a) = \alpha(r_1 a) + \alpha(r_2 a) = (a\alpha)(r_1) + (a\alpha)(r_2)$$

and

$$(a\alpha)(a_1 r) = \alpha((a_1 r)a) = \alpha(a_1(ra)) = a_1\alpha(ra) = a_1((a\alpha)(r)).$$

Thus [7.1] serves to define a scalar multiplication on the left of Hom (R_l, W) by elements of R, and it is an easy matter to show that this scalar multiplication gives the abelian group Hom (R_l, W) the structure of a left R-module.

THEOREM 7.1. *Let R be a ring with identity element e, W a left R-module. Then the left R-module Hom (R_l, W) is isomorphic to W.*

Proof. Consider the mapping f from Hom (R_l, W) to W defined by setting $f(\alpha) = \alpha(e)$ for every homomorphism α from R_l to W. We contend that f is an R-homomorphism; so let α and β be elements of Hom (R_l, W), a any element of R. Then we have

$$f(\alpha + \beta) = (\alpha + \beta)(e) = \alpha(e) + \beta(e) = f(\alpha) + f(\beta)$$

and

$$f(a\alpha) = (a\alpha)(e) = \alpha(ea) = \alpha(ae) = a\alpha(e) = af(\alpha).$$

To show that the homomorphism f is actually the required isomorphism, we notice first that $\alpha \in \operatorname{Ker} f$ if and only if $\alpha(e) = 0$. But if $\alpha(e) = 0$ then $\alpha(r) = \alpha(re) = r\alpha(e) = 0$ for all elements r of R, i.e. α is the zero homomorphism. Thus $\operatorname{Ker} f = \{\zeta\}$, and so f is a monomorphism. Let x be any element of W; if α_x is the mapping from R to W defined by setting $\alpha_x(r) = rx$ for all elements r of R, then $\alpha_x \in \operatorname{Hom}(R_l, W)$ and clearly $f(\alpha_x) = x$. Thus f is an epimorphism.

This completes the proof.

Now let V be any left R-module and consider the group

Hom (V, R_l); the elements of this group are called *linear forms* or *linear functionals* on V. We propose now to give Hom (V, R_l) the structure of a *right R-module*. So let α be any element of Hom (V, R_l), a any element of R and consider the mapping αa from V to R defined by setting

$$(\alpha a)(x) = \alpha(x) a \qquad [7.2]$$

for all elements x of V. Routine verification shows that $\alpha a \in$ Hom (V, R_l) and that under the right multiplication so defined Hom (V, R_l) is a right R-module. (To grasp the reason for the left-right interchange, the reader is invited to check that if we tried defining $a\alpha$ by setting $(a\alpha)(x) = a\alpha(x)$, then $a\alpha$ would not be an R-homomorphism unless R were commutative; if we wrote $(a\alpha)(x) = \alpha(x)a$, we should have $(ab)\alpha = b(a\alpha)$, not $a(b\alpha)$ as we require for a left R-module; but [7.2] gives $\alpha(ab) = (\alpha a)b$, which is the appropriate condition for a right R-module.) When we consider Hom (V, R_l) as a right R-module in this way we call it the *dual module* of V and denote it by V^*. Clearly we may start with a right R-module and form its dual, which will be a left R-module. In particular, if V is a left R-module we may form the dual module of the right R-module V^*; this is again a left R-module, which we denote by V^{**} and call the *bidual* of V.

We digress for a moment to consider the special case in which R is a division ring D and V is a left vector space over D with a finite basis $\{x_1, \ldots, x_n\}$. Then, according to Theorem 5.4, every element x of V can be expressed uniquely in the form $x = a_1 x_1 + \ldots + a_n x_n$ where $a_1, \ldots, a_n$ are elements of D. For $i = 1, \ldots, n$ let ξ_i be the mapping from V to D defined by setting

$$\xi_i(x) = \xi_i(a_1 x_1 + \ldots + a_n x_n) = a_i$$

for each element x of V; it is easily verified that each of these mappings ξ_i is a linear functional on V. We claim that $\{\xi_1, \ldots, \xi_n\}$ is a basis for the right vector space V^* over D. To see this, first let $b_1, \ldots, b_n$ be elements of D such that $\xi_1 b_1 + \ldots + \xi_n b_n = \zeta$, the zero linear functional; then for $j = 1, \ldots, n$ we have

$$0 = (\xi_1 b_1 + \ldots + \xi_n b_n)(x_j) = (\xi_1 b_1)(x_j) + \ldots + (\xi_n b_n)(x_j)$$
$$= \xi_1(x_j) b_1 + \ldots + \xi_n(x_j) b_n = b_j.$$

Thus $\{\xi_1, \ldots, \xi_n\}$ is linearly independent. Next let α be any linear functional on V; for $i = 1, \ldots, n$ let $\alpha(x_i) = b_i$. We claim that $\alpha = \xi_1 b_1 + \ldots + \xi_n b_n$, and verify this by computing

$$\begin{aligned}
\alpha(x) = \alpha(a_1 x_1 + \ldots + a_n x_n) &= a_1 \alpha(x_1) + \ldots + a_n \alpha(x_n) \\
&= a_1 b_1 + \ldots + a_n b_n \\
&= a_1(\xi_1 b_1 + \ldots + \xi_n b_n)(x_1) \\
&\quad + \ldots + a_n(\xi_1 b_1 + \ldots + \xi_n b_n)(x_n) \\
&= (\xi_1 b_1 + \ldots + \xi_n b_n)(x).
\end{aligned}$$

So $\{\xi_1, \ldots, \xi_n\}$ is a generating system for V^*, and hence a basis as asserted. We call $\{\xi_1, \ldots, \xi_n\}$ the basis of V^* *dual* to the basis $\{x_1, \ldots, x_n\}$ of V.

We now return to the general situation, in which R is an arbitrary ring with identity.

Let V_1 and V_2 be two left R-modules, α a module homomorphism from V_1 to V_2. If M is any left R-module we shall show how to define abelian group homomorphisms α^* from Hom (V_2, M) to Hom (V_1, M) and α_* from Hom (M, V_1) to Hom (M, V_2). Namely, if φ is any element of Hom (V_2, M), the mapping $\varphi\alpha$ from V_1 to M is easily seen to be an R-homomorphism and if we set

$$\alpha^*(\varphi) = \varphi\alpha$$

for every element φ of Hom (V_2, M) it turns out that the mapping α^* from Hom (V_2, M) to Hom (V_1, M) so defined is a homomorphism; we call it the homomorphism *induced* by α. Again, if θ is any element of Hom (M, V_1), the mapping $\alpha\theta$ from M to V_2 belongs to Hom (M, V_2) and the mapping α_* from Hom (M, V_1) to Hom (M, V_2) defined by setting

$$\alpha_*(\theta) = \alpha\theta$$

for every element θ of Hom (M, V_1) is a homomorphism, which is also said to be *induced* by α.

The constructions of α^* and α_* which we have just described are special cases of a more general procedure. Suppose V, V', W, W' are left R-modules and that α and β are homomorphisms from V' to V, W to W' respectively; then for each element φ of Hom (V, W) the mapping $\beta\varphi\alpha$ is a homomorphism from V' to W'. This comment allows us to define a mapping from Hom (V, W) to Hom (V', W')

which we denote by Hom (α, β): for each element φ of Hom (V, W) we set

$$(\text{Hom } (\alpha, \beta))(\varphi) = \beta\varphi\alpha.$$

It is a routine matter to verify that Hom (α, β) is an abelian group homomorphism. The mappings α^* and α_* of the previous paragraph become in the present notation simply Hom (α, I_M) and Hom (I_M, α) where I_M is the identity mapping of M. The next two theorems establish fundamental properties of the homomorphisms Hom (α, β).

THEOREM 7.2. *Let* V, V', V'', W, W', W'' *be left R-modules and let* α', α, β, β' *be homomorphisms from* V'' *to* V', V' *to* V, W *to* W', W' *to* W'' *respectively. Then* Hom $(\alpha\alpha', \beta'\beta) = $ Hom (α', β') Hom (α, β).

Proof. Let φ be any element of Hom (V, W). Then we have

$$\begin{aligned}
(\text{Hom } (\alpha\alpha', \beta'\beta))(\varphi) &= (\beta'\beta)\varphi(\alpha\alpha') \\
&= \beta'(\beta\varphi\alpha)\alpha' \\
&= (\text{Hom } (\alpha', \beta'))(\beta\varphi\alpha) \\
&= (\text{Hom } (\alpha', \beta'))((\text{Hom } (\alpha, \beta))(\varphi)).
\end{aligned}$$

Thus Hom $(\alpha\alpha', \beta'\beta) = $ Hom (α', β') Hom (α, β) as required.

COROLLARY. *Let* V_1, V_2, V_3 *be left R-modules and let* α *and* β *be homomorphisms from* V_1 *to* V_2 *and* V_2 *to* V_3 *respectively. Let* M *be any left R-module. Then* $(\beta\alpha)^* = \alpha^*\beta^*$ *and* $(\beta\alpha)_* = \beta_*\alpha_*$.

THEOREM 7.3 *Let* V, V', W, W' *be left R-modules; let* α, α_1, α_2 *be homomorphisms from* V' *to* V *and* β, β_1, β_2 *homomorphisms from* W *to* W'. *Then*

$$\text{Hom } (\alpha, \beta_1 + \beta_2) = \text{Hom } (\alpha, \beta_1) + \text{Hom } (\alpha, \beta_2)$$

and

$$\text{Hom } (\alpha_1 + \alpha_2, \beta) = \text{Hom } (\alpha_1, \beta) + \text{Hom } (\alpha_2, \beta).$$

Proof. Let φ be any element of Hom (V, W). Then we have

$$(\text{Hom } (\alpha, \beta_1 + \beta_2))(\varphi) = (\beta_1 + \beta_2)\varphi\alpha$$

while

$$\begin{aligned}
(\text{Hom } (\alpha, \beta_1) + \text{Hom } (\alpha, \beta_2))(\varphi) &= (\text{Hom } (\alpha, \beta_1))(\varphi) + (\text{Hom } (\alpha, \beta_2))(\varphi) \\
&= \beta_1\varphi\alpha + \beta_2\varphi\alpha.
\end{aligned}$$

Now if x' is any element of V' we have

$$[(\beta_1 + \beta_2)\,\varphi\alpha]\,(x') = (\beta_1 + \beta_2)\,(\varphi\alpha(x'))$$
$$= \beta_1(\varphi\alpha(x')) + \beta_2(\varphi\alpha(x'))$$
$$= \beta_1\varphi\alpha(x') + \beta_2\varphi\alpha(x') = [\beta_1\varphi\alpha + \beta_2\varphi\alpha]\,(x').$$

Hence $\mathrm{Hom}\,(\alpha, \beta_1 + \beta_2) = \mathrm{Hom}\,(\alpha, \beta_1) + \mathrm{Hom}\,(\alpha, \beta_2)$.
The other result follows similarly.

If R is a commutative ring and the various groups of homomorphisms are given the R-module structure defined earlier, the various mappings $\mathrm{Hom}\,(\alpha, \beta)$, α^*, α_* are all R-module homomorphisms.

We now examine what happens to short exact sequences of left R-modules when we form the homomorphism groups of the terms from and to some fixed left R-module. In the course of our discussions various zero homomorphisms will occur; if it is necessary to emphasise that we are dealing (for example) with the zero homomorphism from V to W we shall denote it by ζ_{VW}, but usually we shall denote all zero homomorphisms simply by ζ since the context in most cases will make it clear which particular zero homomorphism is meant.

THEOREM 7.4. *Let*

$$0 \to V' \xrightarrow{\alpha} V \xrightarrow{\beta} V'' \to 0 \qquad [7.3]$$

be a short exact sequence of left R-modules, M a left R-module. Then the sequences

$$0 \to \mathrm{Hom}\,(V'', M) \xrightarrow{\beta^*} \mathrm{Hom}\,(V, M) \xrightarrow{\alpha^*} \mathrm{Hom}\,(V', M) \qquad [7.4]$$

and

$$0 \to \mathrm{Hom}\,(M, V') \xrightarrow{\alpha_*} \mathrm{Hom}\,(M, V) \xrightarrow{\beta_*} \mathrm{Hom}\,(M, V'') \qquad [7.5]$$

are exact.

Proof. (1) To show that [7.4] is exact we must show that (a) β^* is a monomorphism, (b) $\mathrm{Im}\,\beta^* \subseteq \mathrm{Ker}\,\alpha^*$ and (c) $\mathrm{Ker}\,\alpha^* \subseteq \mathrm{Im}\,\beta^*$.

(a) Let θ'' be an element of $\mathrm{Ker}\,\beta^*$; then $\theta''\beta = \beta^*(\theta'') = \zeta_{VM}$. If x'' is any element of V'' then there is an element x of V such that $x'' = \beta(x)$. It follows that $\theta''(x'') = \theta''(\beta(x)) = 0$. Thus $\theta'' = \zeta_{V''M}$ and hence β^* is a monomorphism.

(b) Now let θ be any element of $\mathrm{Im}\,\beta^*$; thus there is an element θ''

of Hom (V'', M) such that $\theta = \beta^*(\theta'') = \theta''\beta$. Then $\alpha^*(\theta) = \theta\alpha = (\theta''\beta)\alpha = \theta''(\beta\alpha) = \theta''\zeta = \zeta$. Hence Im $\beta^* \subseteq$ Ker α^*. (We have $\beta\alpha = \zeta$ since [7.3] is exact.)

(c) Next let θ be any element of Ker α^*. Then for every element x' of V' we have $\theta\alpha(x') = [\alpha^*(\theta)](x') = 0$. Consider now the mapping θ'' from V'' to M defined as follows: for each element x'' of V'' choose an element x from $\beta^{-1}(x'')$—which is non-empty since β is an epimorphism—and form $\theta(x)$. This element depends only on x'', not on the choice of x; for if x and x_1 are elements of $\beta^{-1}(x'')$ we have $x - x_1 \in$ Ker $\beta =$ Im α and hence $\theta(x - x_1) = 0$. So we may properly define $\theta''(x'')$ to be $\theta(x)$. It is easy to verify that in fact $\theta'' \in$ Hom (V'', M), and clearly $\beta^*(\theta'') = \theta''\beta = \theta$. Hence Ker $\alpha^* \subseteq$ Im β^*.

Thus [7.4] is exact.

(2) To show that [7.5] is exact we must establish that (a) α_* is a monomorphism, (b) Im $\alpha_* \subseteq$ Ker β_* and (c) Ker $\beta_* \subseteq$ Im α_*.

(a) Let θ' be an element of Ker α_* so that $\alpha\theta' = \alpha_*(\theta') = \zeta_{MV}$. Thus if m is any element of M, we have $\alpha(\theta'(m)) = 0$. Since [7.3] is exact, α is a monomorphism and hence $\theta'(m) = 0$. Thus $\theta' = \zeta_{MV'}$ and hence α_* is a monomorphism.

(b) Let θ be an element of Im α_*; thus there is an element θ' of Hom (M, V') such that $\theta = \alpha_*(\theta')$. Then we have $\beta_*(\theta) = \beta_*(\alpha_*(\theta')) = \beta(\alpha\theta') = (\beta\alpha)\theta' = \zeta$. Hence Im $\alpha_* \subseteq$ Ker β_*. (We use here again the fact that $\beta\alpha = \zeta$, which follows from the exactness of [7.3].)

(c) Finally let θ be any element of Ker β_*; then for every element m of M we have $\beta(\theta(m)) = \beta\theta(m) = [\beta_*(\theta)](m) = 0$, i.e. $\theta(m) \in$ Kerβ. Since [7.3] is exact, it follows that $\theta(m) \in$ Im α. Thus there is an element x' of V' such that $\theta(m) = \alpha(x')$; and x' is unique since α is a monomorphism. Set $\theta'(m) = x'$; in this way we define a mapping θ' from M to V', which is clearly a homomorphism. Since $\theta = \alpha\theta' = \alpha_*(\theta')$ it follows that $\theta \in$ Im α_*. Hence Ker $\beta_* \subseteq$ Im α_*.

Thus [7.5] is exact.

Given a short exact sequence [7.3] of left R-modules, we are naturally interested in the sequence of homomorphism groups

$$0 \to \text{Hom } (V'', M) \xrightarrow{\beta^*} \text{Hom } (V, M) \xrightarrow{\alpha^*} \text{Hom } (V', M) \to 0. \qquad [7.6]$$

In Theorem 9.8 we shall obtain necessary and sufficient conditions on the module M in order that [7.6] be exact for every short exact sequence [7.3]; in Theorem 9.6 we carry out a similar programme in connexion with the sequence

$$0 \to \operatorname{Hom}(M, V') \overset{\alpha*}{\to} \operatorname{Hom}(M, V) \overset{\beta*}{\to} \operatorname{Hom}(M, V'') \to 0. \quad [7.7]$$

In Theorem 8.7 we obtain an important class of sequences [7.3] such that [7.6] and [7.7] are exact for every module M.

If V is any left R-module, the set $\operatorname{Hom}_R(V, V)$ of R-endomorphisms of V admits not only the addition operation described at the beginning of this section but also the composition operation. As in Example 6 of §2 we may verify that $\operatorname{Hom}_R(V, V)$ forms a ring with these two operations as addition and multiplication respectively. There is an important special case in which we can give a 'concrete' representation of this ring.

THEOREM 7.5. *Let V be a left vector space over a division ring D. If V has a basis consisting of n elements then $\operatorname{Hom}_D(V, V)$ is isomorphic to the ring $M_n(D^{op})$ of $n \times n$ matrices with coefficients in the ring opposite to D.*

Proof. Let $\{x_1, \ldots, x_n\}$ be a basis for V.

If α is any element of $\operatorname{Hom}_D(V, V)$ then for $j = 1, \ldots, n$ we may express $\alpha(x_j)$ uniquely in the form $\alpha(x_j) = \sum_{i=1}^{n} a_{ij} x_i$. Thus we may define a mapping φ from $\operatorname{Hom}_D(V, V)$ to $M_n(D^{op})$ by setting $\varphi(\alpha) = [a_{ij}]$ for each D-endomorphism α of V. It is almost immediate that φ is bijective. To see that it is an isomorphism, let β be another element of $\operatorname{Hom}_D(V, V)$ and set $\varphi(\beta) = [b_{ij}]$. Then for $j = 1, \ldots, n$ we have

$$(\alpha + \beta)(x_j) = \alpha(x_j) + \beta(x_j)$$

$$= \sum_{i=1}^{n} a_{ij} x_i + \sum_{i=1}^{n} b_{ij} x_i$$

$$= \sum_{i=1}^{n} (a_{ij} + b_{ij}) x_i;$$

so $\varphi(\alpha + \beta) = [a_{ij} + b_{ij}] = [a_{ij}] + [b_{ij}] = \varphi(\alpha) + \varphi(\beta)$. Further, for $j = 1, \ldots, n$ we have

$$(\beta\alpha)(x_j) = \beta(\alpha(x_j))$$

$$= \beta\left(\sum_{k=1}^{n} a_{kj} x_k\right)$$

$$= \sum_{k=1}^{n} a_{kj}\beta(x_k)$$

$$= \sum_{k=1}^{n} a_{kj} \left(\sum_{i=1}^{n} b_{ik}x_i \right)$$

$$= \sum_{i=1}^{n} \left(\sum_{k=1}^{n} a_{kj}b_{ik} \right) x_i.$$

Hence $\varphi(\beta\alpha)$ is the matrix with (i,j)-th element $\sum_{k=1}^{n} a_{kj}b_{ik} = \sum_{k=1}^{n} b_{ik}*a_{kj}$ (where $*$ denotes the multiplication operation in D^{op}); this shows that $\varphi(\beta\alpha) = \varphi(\beta)*\varphi(\alpha)$. So φ is an isomorphism, as asserted.

§8. Direct Products and Sums

Let R be a ring (with identity), $(V_k)_{k \in K}$ a family of (unitary) left R-modules. (There is no sinister reason for using K to denote the index set instead of the more familiar I; we shall have to consider homomorphisms which are conventionally denoted by ι, and we wish to avoid confusion between mappings ι and indices i—the notation ι_i would be a particularly unhappy one.)

We form the product $P = \prod_{k \in K} V_k$ of the family $(V_k)_{k \in K}$; if x is any element of P we write $x_k = x(k)$ for each index k in K and denote x by $(x_k)_{k \in K}$ or simply by (x_k). For each index k we have the kth projection π_k which maps P onto V_k according to the prescription $\pi_k(x) = x_k$ for every element x of P.

We define an addition operation in P as follows. If $x = (x_k)$ and $y = (y_k)$ are any elements of P we set $x + y = (z_k)$ where $z_k = x_k + y_k \, (\in V_k)$ for every index k in K. Next we define a scalar multiplication on the left of P by elements of R. If $x = (x_k)$ is any element of P and a is any element of R, set $ax = (t_k)$ where $t_k = ax_k \, (\in V_k)$ for each k in K. It is an easy matter to verify that P is a left R-module under the addition and scalar multiplication we have just defined, and that the mappings π_k are all homomorphisms. We call P the *direct product module* of the family $(V_k)_{k \in K}$ and the homomorphisms π_k are called the *canonical projection epimorphisms* from P onto the modules V_k. For each index j in K we define a mapping ι_j from V_j to P by setting for each element x_j of V_j

$$\iota_j(x_j) = y = (y_k)_{k \in K}$$

where $y_j = x_j$ and $y_k = 0$ for $k \neq j$. Clearly each of these mappings is injective, and it is easy to verify that they are all homomorphisms; we call ι_k the *canonical injection monomorphism* from V_k to P. It follows at once from the definitions that for each index k in K we have $\pi_k \iota_k = I_k$ (the identity mapping of V_k) while for each pair of distinct indices k, l in K we have $\pi_k \iota_l = \zeta_{lk}$ (the zero homomorphism from V_l to V_k).

Let (V_k), (W_k) be two families of left R-modules indexed by the same set K. For each index k in K let φ_k be a homomorphism from V_k to W_k. Then we can define a mapping φ from $\prod_{k \in K} V_k$ to $\prod_{k \in K} W_k$ by setting $\varphi(x) = (\varphi_k(x_k))$ for each element $x = (x_k)$ of $\prod_{k \in K} V_k$. It is easily verified that φ is a homomorphism; we say that φ is induced by the family (φ_k).

THEOREM 8.1. *Let (V'_k), (V_k), (V''_k) be three families of left R-modules, all indexed by the same set K. Let (φ_k), (θ_k) be families of homomorphisms also indexed by K such that for each index k in K the sequence*

$$V'_k \xrightarrow{\;\varphi_k\;} V_k \xrightarrow{\;\theta_k\;} V''_k$$

is exact. Then the sequence

$$\prod_{k \in K} V'_k \xrightarrow{\;\varphi_k\;} \prod_{k \in K} V_k \xrightarrow{\;\theta_k\;} \prod_{k \in K} V''_k$$

is also exact, where φ and θ are the homomorphisms induced by (φ_k) and (θ_k) respectively.

Proof. (1) Let $x = (x_k)$ be any element of Im φ, say $x = \varphi(x')$, where $x' = (x'_k) \subset \prod_{k \in K} V'_k$; then for each index k in K we have $x_k = \varphi_k(x'_k)$. It follows that $\theta(x) = (\theta_k(x_k)) = (\theta_k \varphi_k(x'_k)) = 0$ since Im $\varphi_k = $ Ker θ_k for each index k.

So Im $\varphi \subseteq$ Ker θ.

(2) Conversely, suppose $x = (x_k) \in$ Ker θ. Then $\theta(x) = (\theta_k(x_k)) = 0$ and hence $\theta_k(x_k) = 0$ for each index k in K. Since Ker $\theta_k = $ Im φ_k for each index k it follows that for each index k there is an element x'_k of V'_k such that $x_k = \varphi_k(x'_k)$. If we set $x' = (x'_k)$, then x' is an element of $\prod_{k \in K} V'_k$ and $x = \varphi(x')$.

So Ker $\theta \subseteq$ Im φ.

This completes the proof.

Let $(V_k)_{k \in K}$ be a family of left R-modules. Let $(P', (\pi'_k))$ be a pair consisting of a left R-module P' and a family $(\pi'_k)_{k \in K}$ of homomorphisms such that for each index k in K the homomorphism π'_k maps P' into V_k. We say that $(P', (\pi'_k))$ is *couniversal* for homomorphisms to the family (V_k) if for every left R-module V and every family (φ_k) of homomorphisms from V to the modules V_k, there exists a unique homomorphism α from V to P' such that the diagrams

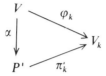

are commutative for every index k in K. (We say in this case that the homomorphisms φ_k can be uniquely factored through P'.)

THEOREM 8.2 *If $(V_k)_{k \in K}$ is any family of left R-modules, then the pair $(P, (\pi_k))$ consisting of the direct product module P of the family (V_k) and the family of canonical projection epimorphisms is couniversal for homomorphisms to the family (V_k).*

Proof. Let V be any left R-module, (φ_k) a family of homomorphisms from V to the modules V_k.

We define a mapping α from V to P by setting for each element x of V

$$\alpha(x) = (y_k)_{k \in K}$$

where $y_k = \varphi_k(x) \in V_k$ for each index k in K. Then α is easily seen to be a homomorphism, and for each index k we have

$$\pi_k \alpha(x) = y_k = \varphi_k(x)$$

for all elements x of V. So $\pi_k \alpha = \varphi_k$ for all indices k in K.

We claim that α is the only homomorphism from V to P with this property. So let α' be another homomorphism from V to P such that $\pi_k \alpha' = \varphi_k$ for every index k. Let x be any element of V. Then for each index k in K we have

$$(\alpha(x))(k) = \pi_k(\alpha(x)) = \varphi_k(x) = \pi_k(\alpha'(x)) = (\alpha'(x))(k).$$

So $\alpha(x) = \alpha'(x)$ for every element x of V and hence $\alpha = \alpha'$.

Thus $(P,(\pi_k))$ is couniversal for homomorphisms to the family (V_k).

Let V be any left R-module, (φ_k) a family of homomorphisms from V to (V_k). Then, as we have just seen, there exists a unique homomorphism α from V to P such that $\pi_k\alpha = \varphi_k$ for all indices k in K. If this homomorphism is actually an isomorphism then the homomorphisms φ_k are all surjective and we say that the family (φ_k) yields a *projective representation* of V as a direct product of the family (V_k). In this situation the inverse α^{-1} of the isomorphism α is an isomorphism from P to V and so for each index k in K we can define a homomorphism η_k from V_k to V by setting $\eta_k = \alpha^{-1}\iota_k$, where ι_k is the canonical injection monomorphism from V_k to P. These homomorphisms η_k are all injective, and for each index k in K we have $\varphi_k\eta_k = (\pi_k\alpha)(\alpha^{-1}\iota_k) = \pi_k\iota_k = I_k$, while for each pair of distinct indices k, l in K we have $\varphi_k\eta_l = (\pi_k\alpha)(\alpha^{-1}\iota_l) = \pi_k\iota_l = \zeta_{lk}$. Thus the families (φ_k) and (η_k) stand in the same relation to V as the families (π_k) and (ι_k) to P.

Still let $(V_k)_{k \in K}$ be a family of left R-modules, P their direct product module. Let S be the subset of P consisting of those elements $x = (x_k)$ such that the coordinates x_k are non-zero for at most finitely many indices k. It is easily verified that S is a submodule of P, that the canonical injection monomorphism ι_k actually maps V_k into S and that the restriction to S of each canonical projection epimorphism π_k is an epimorphism from S onto V_k. We call S the *external direct sum* of the family (V_k) and denote it by $\bigoplus_{k \in K} V_k$.

If x is any element of S the elements $\iota_k\pi_k(x)$ of S are non-zero for at most finitely many indices k in K. Thus we may form the sum $\sum_{k \in K} \iota_k\pi_k(x)$, and it is clear that in fact $\sum_{k \in K} \iota_k\pi_k(x) = x$.

Let (V_k), (W_k) be two families of left R-modules indexed by the same set K and for each index k in K let φ_k be a homomorphism from V_k to W_k. Then we can define a mapping φ from $\bigoplus_{k \in K} V_k$ to $\prod_{k \in K} W_k$ by setting $\varphi(x) = (\varphi_k(x_k))$ for each element $x = (x_k)$ of $\bigoplus V_k$. Since $x_k = 0$ for all but a finite set of indices k, $\varphi_k(x_k)$ is non-zero for at most finitely many indices k; hence φ actually maps $\bigoplus V_k$ into $\bigoplus W_k$. It is a trivial matter to verify that φ is a homomorphism; we say that φ is induced by the family (φ_k).

The analogue of Theorem 8.1 is easy to establish. We leave the proof to the reader.

THEOREM 8.3. *Let (V_k'), (V_k), (V_k'') be three families of left R-modules, all indexed by the same set K. Let (φ_k), (θ_k) be families of homomorphisms also indexed by K such that for each index k in K the sequence*

$$V_k' \xrightarrow{\varphi_k} V_k \xrightarrow{\theta_k} V_k''$$

is exact. Then the sequence

$$\oplus_{k \in K} V_k' \xrightarrow{\varphi} \oplus_{k \in K} V_k \xrightarrow{\theta} \oplus_{k \in K} V_k''$$

is also exact, where φ and θ are the homomorphisms induced by (φ_k) and (θ_k) respectively.

Let S' be a left R-module, $(\iota_k')_{k \in K}$ a family of homomorphisms such that for each index k in K the homomorphism ι_k' maps V_k into S'. We say that the pair $(S', (\iota_k'))$ is *universal* for homomorphisms from the family (V_k) if for every left R-module V and every family $(\eta_k)_{k \in K}$ of homomorphisms from the modules V_k to V there exists a unique homomorphism β from S' to V such that the diagrams

are commutative for every index k in K. (We say that the homomorphisms can be uniquely factored through S'.)

THEOREM 8.4. *If $(V_k)_{k \in K}$ is any family of left R-modules, then the pair $(S, (\iota_k))$ consisting of the external direct sum of the family (V_k) and the family of canonical injection monomorphisms is universal for homomorphisms from the family (V_k).*

Proof. Let V be any left R-module, (η_k) a family of homomorphisms from the modules V_k to V.

We define a mapping β from S to V by setting

$$\beta(x) = \sum_{k \in K} \eta_k(x_k) = \sum_{k \in K} \eta_k \pi_k(x)$$

for each element $x = (x_k)$ of S. (These sums are well-defined since only finitely many of the elements x_k are non-zero). Then β is easily seen to be a homomorphism. For each index l in K and each element x_l of V_l we have

$$\beta \iota_l(x_l) = \sum_{k \in K} \eta_k \pi_k \iota_l(x_l) = \eta_l(x_l)$$

since $\pi_k \iota_l = \zeta_{lk}$ when $k \neq l$ and $\pi_l \iota_l = I_l$; thus $\beta \iota_l = \eta_l$.

We claim that β is the only homomorphism from S to V with this property. So let β' be another homomorphism from S to V such that $\beta' \iota_k = \eta_k$ for all indices k in K. If x is any element of S we have

$$\beta(x) = \beta(\sum \iota_k \pi_k(x)) = \sum \beta \iota_k \pi_k(x) = \sum \eta_k \pi_k(x)$$

and similarly

$$\beta'(x) = \sum \eta_k \pi_k(x).$$

Hence $\beta = \beta'$ as asserted.

Thus $(S, (\iota_k))$ is universal for homomorphisms from (V_k).

Let V be any left R-module, (η_k) a family of homomorphisms from (V_k) to V. As we have just proved in Theorem 8.4, there exists a unique homomorphism β from S to V such that $\beta \iota_k = \eta_k$ for every index k in K. If this homomorphism is actually an isomorphism, then the homomorphisms η_k are all injective and we say that the family (η_k) yields an *injective representation* of V as a direct sum of the family (V_k). The inverse β^{-1} of β is an isomorphism from V to S and so for every index k in K we have a homomorphism $\varphi_k = \pi_k \beta^{-1}$ from V to V_k; these homomorphisms are all surjective. Now for each index k in K we have $\varphi_k \eta_k = (\pi_k \beta^{-1})(\beta \iota_k) = \pi_k \iota_k = I_k$, and for every pair of distinct indices k, l in K we have $\varphi_k \eta_l = (\pi_k \beta^{-1})(\beta \iota_l) = \pi_k \iota_l = \zeta_{lk}$. Further, for every element x of V we have

$$\sum \eta_k \varphi_k(x) = \sum \beta \iota_k \pi_k(\beta^{-1}(x)) = \beta(\sum \iota_k \pi_k(\beta^{-1}(x))) = \beta(\beta^{-1}(x)) = x.$$

Hence in this situation the families (φ_k) and (η_k) play the same roles for V as the families (π_k) and (ι_k) do for S.

Let V be a left R-module, $(W_k)_{k \in K}$ a family of submodules of V. In §5 we defined the sum $W = \sum_{k \in K} W_k$ of this family and showed that it is the submodule of V consisting of all elements of V of the form $\sum\limits_{k \in K} w_k$ where $w_k \in W_k$ for each index k in K and only finitely many of the elements w_k are non-zero. For each index k in K the inclusion monomorphism from W_k to V, which we shall denote here by η_k, actually maps W_k into W. Using now the universal property of the external direct sum $S = \oplus_{k \in K} W_k$ it follows that there exists a unique homomorphism β from S to W such that $\beta\iota_k = \eta_k$ for every index k in K. If β is an isomorphism from S to W, so that the family (η_k) yields an injective representation of W as a direct sum of the family (W_k), we say that W is the *internal direct sum* of the family (W_k) and that the sum $\sum\limits_{k \in K} W_k$ is *direct*. We now derive necessary and sufficient conditions for the sum of a family of submodules to be direct.

THEOREM 8.5. *Let* $(W_k)_{k \in K}$ *be a family of submodules of a left R-module V. Then the following conditions are equivalent*:

(a) *the sum* $W = \sum\limits_{k \in K} W_k$ *is direct*;

(b) *every element of W can be expressed uniquely in the form* $\sum\limits_{k \in K} w_k$ *where $w_k \in W_k$ for each index k in K and only finitely many of the elements w_k are non-zero*;

(c) *for each index k_0 in K the intersection* $W_{k_0} \cap \sum\limits_{k \in K, k \neq k_0} W_k$ *consists of the zero element alone*.

Proof. (1) First we prove that (a) implies (b). So we suppose that the sum $W = \sum W_k$ is direct.

There is thus a unique isomorphism β from $S = \oplus W_k$ onto W; according to Theorem 8.4 β is defined by setting, for each element x of S,

$$\beta(x) = \sum\limits_{k \in K} \eta_k \pi_k(x). \qquad [8.1]$$

Suppose now $w = \sum w_k = \sum w_k'$ where w_k, $w_k' \in W_k$ for each index k and only finitely many of the elements w_k, w_k' are non-zero. Consider the elements $x = (w_k)$ and $x' = (w_k')$ of S. Then

$$\beta(x) = \sum \eta_k \pi_k(x) = \sum \eta_k(w_k) = \sum w_k = w$$

and similarly $\beta(x') = w$. Since β is an isomorphism it follows that $x = x'$ and hence $w_k = w_k'$ for each index k, as required.

(2) Next suppose that every element of W can be expressed uniquely in the form described.

Let k_0 be any index in K, w any element of $W_{k_0} \cap \sum\limits_{k \neq k_0} W_k$. Since $w \in W_{k_0}$, it can be expressed in the form $w = \sum\limits_{k \in K} v_k$ where $v_{k_0} = w$ and $v_k = 0$ for $k \neq k_0$; since $w \in \sum\limits_{k \neq k_0} W_k$ it can be expressed in the form $w = \sum\limits_{k \in K} v'_k$ where $v'_{k_0} = 0$, and for $k \neq k_0$ we have $v'_k \in W_k$ (and only finitely many of the elements v'_k are non-zero). It now follows from the uniqueness hypothesis that $w = v_{k_0} = v'_{k_0} = 0$.

Thus (b) implies (c).

(3) Finally suppose that for each index k_0 in K we have $W_{k_0} \cap \sum\limits_{k \neq k_0} W_k = \{0\}$ and consider the homomorphism β defined by the equation [8.1].

Clearly β is an epimorphism; as we saw in (1) above, if $w = \sum w_k$ is any element of W then $w = \beta(x)$ where $x = (w_k)$.

To show that β is a monomorphism, let x be any element of S such that $\beta(x) = \sum \eta_k \pi_k(x) = 0$. If x is non-zero there is at least one index k_0 in K such that $\pi_{k_0}(x) \neq 0$ and hence $\eta_{k_0} \pi_{k_0}(x)$ is a non-zero element of W_{k_0}. But we also have $\eta_{k_0} \pi_{k_0}(x) = - \sum\limits_{k \neq k_0} \eta_k \pi_k(x) \in \sum\limits_{k \neq k_0} W_k$. This contradicts our hypothesis that $W_{k_0} \cap \sum\limits_{k \neq k_0} W_k = \{0\}$. Hence $x = 0$ and β is a monomorphism.

Thus the sum $W = \sum W_k$ is direct; so (c) implies (a).

This completes the proof.

We now specialise to the case where $(V_k)_{k \in K}$ is a finite family of left R-modules. In this case the direct product $P = \prod\limits_{k \in K} V_k$ and the external direct sum $S = \oplus_{k \in K} V_k$ coincide. If $K = [1, n]$ we denote the module $P = S$ by

$$V_1 \oplus V_2 \oplus \ldots \oplus V_n$$

and we may write the elements of this module in the form $x = (x_1, \ldots, x_n)$ where $x_k \in V_k$ $(k = 1, \ldots, n)$. Suppose that the family (η_k) of monomorphisms from (V_k) to a module V yields an injective representation of V as a direct sum of the family (V_k). Then, as we have seen (just after Theorem 8.4.), there is a family (φ_k) of epimorphisms from V onto the modules of the family (V_k). It is easy to verify that this family (φ_k) provides a projective representation of V as a direct product of the family (V_k). We can show similarly that if a family (φ_k) of epimorphisms from V onto the modules of the family (V_k) provides a projective representation of V as a direct product of the family (V_k) then the family (η_k) of monomorphisms from (V_k) into V described in our discussion after Theorem 8.2 provides an injective representation of V as a direct sum of the family (V_k).

Let V be a left R-module, W a submodule. Then W is said to be a *direct summand* of V if there exists a submodule W' of V such that V is the internal direct sum of W and W'. Every such submodule W' is called a *supplement* of W, and we say that W and W' are *supplementary submodules* of V. It follows immediately from Theorem 8.5 that the submodules W and W' are supplementary if and only if $W + W' = V$ and $W \cap W' = 0$. In this situation every element x of V can be expressed uniquely in the form $x = w + w'$ with w in W and w' in W'. We notice also that according to Theorem 6.4 the factor module $V/W = (W + W')/W$ is isomorphic to $W'/(W \cap W')$, and hence to W' (since $W \cap W' = 0$); similarly, of course, V/W' is isomorphic to W.

A short exact sequence

$$0 \to V' \xrightarrow{\alpha} V \xrightarrow{\beta} V'' \to 0 \qquad [8.2]$$

is called a *split exact sequence* if the submodule Im $\alpha = \alpha(V')$ is a direct summand of V. We now establish alternative criteria for a short exact sequence to be split.

THEOREM 8.6. *The following conditions for the short exact sequence* [8.2] *are equivalent*:

(a) *the sequence is split*;
(b) *there is an epimorphism α' from V onto V' such that $\alpha'\alpha = I_{V'}$*;
(c) *there is a monomorphism β'' from V'' to V such that $\beta\beta'' = I_{V''}$*.

Proof. (1) Suppose the sequence [8.2] is split. Then there exists a submodule W of V such that V is the internal direct sum of Im α and W, and hence every element x of V can be expressed uniquely in the form $x = v + w$ where $v \in $ Im α and $w \in W$. Since α is a monomorphism, there is a unique element x' of V' such that $v = \alpha(x')$; set $\alpha'(x) = x'$. The mapping α' from V to V' so defined is easily seen to be the epimorphism required in condition (b).

(2) Suppose condition (b) is satisfied.

Let x'' be any element of V''; since [8.2] is exact, β is an epimorphism and hence there is an element x of V such that $\beta(x) = x''$. We claim that the element $x - \alpha\alpha'(x)$ depends only on x'' and not on the choice of x. To see this, let x_1 be another element of V such that $\beta(x_1) = x''$. Then $x - x_1 \in \operatorname{Ker} \beta = \operatorname{Im} \alpha$; say $x - x_1 = \alpha(x')$ where $x' \in V'$. It follows that $\alpha\alpha'(x - x_1) = \alpha\alpha'\alpha(x') = \alpha(x') = x - x_1$, whence $x - \alpha\alpha'(x) = x_1 - \alpha\alpha'(x_1)$.

It is now easy to see that the mapping β'' from V'' to V defined by setting $\beta''(x'') = x - \alpha\alpha'(x)$ is the monomorphism required for condition (c).

(3) Finally suppose condition (c) is satisfied. We shall show that V is the internal direct sum of $\operatorname{Im} \alpha$ and $\operatorname{Im} \beta''$. If $x \in \operatorname{Im} \alpha \cap \operatorname{Im} \beta''$ there exist elements x' of V' and x'' of V'' such that $x = \alpha(x') = \beta''(x'')$. Then $0 = \beta\alpha(x') = \beta\beta''(x'') = x''$; hence $x = 0$ and so $\operatorname{Im} \alpha \cap \operatorname{Im} \beta'' = 0$. Next, if x is any element of V we see at once that $x - \beta''\beta(x) \in \operatorname{Ker} \beta = \operatorname{Im} \alpha$; so $x \in \operatorname{Im} \alpha + \operatorname{Im} \beta''$.

Thus V is the internal direct sum of $\operatorname{Im} \alpha$ and $\operatorname{Im} \beta''$, and condition (a) is satisfied.

It is easy to see that when [8.2] splits we have a commutative diagram

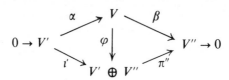

where $\imath'$ is the monomorphism given by $\imath'(x') = (x', 0)$ for all elements x' in V', π'' is the epimorphism given by $\pi''(x', x'') = x''$ for all elements (x', x'') in $V' \oplus V''$ and φ is the isomorphism given by $\varphi(x) = (\alpha'(x), \beta(x))$ for all elements x in V.

Now that we have introduced the idea of a split exact sequence we can redeem one of the promises which we made at the end of §7.

THEOREM 8.7. *Let* [8.2] *be a split exact sequence of left R-modules. Then for every left R-module M the sequences*

$$0 \to \operatorname{Hom}(V'', M) \xrightarrow{\beta^*} \operatorname{Hom}(V, M) \xrightarrow{\alpha^*} \operatorname{Hom}(V', M) \to 0 \quad [8.3]$$

and

$$0 \to \mathrm{Hom}\,(M, V') \xrightarrow{\alpha_*} \mathrm{Hom}\,(M, V) \xrightarrow{\beta_*} \mathrm{Hom}\,(M, V'') \to 0 \quad [8.4]$$

are split exact.

Proof. Since [8.2] is split, it follows from Theorem 8.6 that there exists an epimorphism α' from V onto V' such that $\alpha'\alpha = I_{V'}$ and a monomorphism β'' from V'' to V such that $\beta\beta'' = I_{V''}$.

(1) To show that [8.3] is exact all that remains to be shown is that α^* is surjective (cf. Theorem 7.4).

So let θ' be any element of Hom (V', M). If we set $\theta = \theta'\alpha'$ then $\theta \in \mathrm{Hom}\,(V, M)$ and we have $\alpha^*(\theta) = \theta\alpha = (\theta'\alpha')\,\alpha = \theta'(\alpha'\alpha) = \theta'$, i.e. α^* is surjective as required.

Since $\alpha'\alpha = I_{V'}$ we have $\alpha^*(\alpha')^* = (\alpha'\alpha)^* = I_{V'}^*$, from which it follows that [8.3] is split.

(2) To show that [8.4] is exact we have only to show that β_* is surjective.

So let θ'' be any element of Hom (M, V''). If this time we set $\theta = \beta''\theta''$ then $\theta \in \mathrm{Hom}\,(M, V)$ and $\beta_*(\theta) = \beta\theta = \theta''$; so β_* is surjective.

The fact that [8.4] is split follows from the fact that $\beta\beta'' = I_{V''}$.

§9. Free, Projective and Injective Modules

Let R be a ring with identity element e, E any set. Consider the family $(V_x)_{x \in E}$ of left R-modules indexed by E in which each member V_x is the module R_l, i.e. the ring itself considered as left R-module. We form the external direct sum of this family and denote it by $F(E)$. According to the definition of the external direct sum of a family of modules, $F(E)$ is the subset of Map (E, R) consisting of those mappings α from E to R such that $\alpha(x)$ is non-zero for at most finitely many elements x of E; of course if E is a finite set we have $F(E) = $ Map (E, R). We recall the definitions of the addition and scalar multiplication in $F(E)$. If α_1 and α_2 are elements of $F(E)$, we define $\alpha_1 + \alpha_2$ by setting

$$(\alpha_1 + \alpha_2)(x) = \alpha_1(x) + \alpha_2(x) \qquad [9.1]$$

for each element x of E; if a is any element of R and α any element of $F(E)$, we set

$$(a\alpha)(x) = a\alpha(x)$$

for each element x of E.

The module $F(E)$ is called the *free R-module based on E*.

Consider next the mapping φ from E to $F(E)$ defined by setting $\varphi(x) = \iota_x(e)$ for every element x of E, where ι_x is the canonical injection monomorphism from $V_x = R_1$ to $F(E)$. We call φ the *natural mapping* from E to $F(E)$.

If α is any element of $F(E)$ then for each element x of E we write a_x instead of $\pi_x(\alpha) = \alpha(x)$. Then we have

$$\alpha = \sum_{x \in E} \iota_x \pi_x(\alpha)$$

$$= \sum_{x \in E} \iota_x(\pi_x(\alpha) . e)$$

$$= \sum_{x \in E} \pi_x(\alpha) \iota_x(e) = \sum_{x \in E} a_x \varphi(x).$$

If we identify E with its image under φ in $F(E)$ we have expressed α as a 'formal sum'

$$\alpha = \sum_{x \in E} a_x x \qquad [9.2]$$

of elements of E with coefficients in R. The expression [9.2] for α is plainly unique.

We say that a pair (G, θ) consisting of a left R-module G and a mapping θ from E to G is *universal* for mappings from E to left R-modules if for every mapping ψ from E to a left R-module V there exists a unique R-module homomorphism v from G to V such that the diagram

is commutative.

THEOREM 9.1. *Let E be any set. The pair $(F(E), \varphi)$ consisting of the free R-module based on E and the natural mapping from E to F(E) is universal for mappings from E to left R-modules.*

Proof. Throughout the proof we use F as an abbreviation for $F(E)$.

Let V be any left R-module, ψ a mapping from E to V. Consider the mapping v from F to V defined by setting for each element α of F

$$v(\alpha) = \sum_{x \in E} \alpha(x)\psi(x).$$

(If we express α as a formal sum, $\alpha = \sum_{x \in E} \alpha_x x$, this definition becomes

$$v(\sum a_x x) = \sum a_x \psi(x).)$$

Then it is easy to verify that v is a homomorphism. Further, $v\varphi = \psi$; for if t is any element of E, we have

$$v(\varphi(t)) = \sum_{x \in E} [\varphi(t)](x)\psi(x) = \psi(t).$$

The homomorphism v is plainly unique.

So $(F(E), \varphi)$ is universal for mappings from E to left R-modules.

A left R-module G is said to be *free* if there exists a set E such that G is isomorphic to $F(E)$, the free left R-module based on E. The next theorem gives a useful criterion for a module to be free.

THEOREM 9.2. *A left R-module is free if and only if it has a basis.*

Proof. (1) Let G be a left R-module with basis X. We shall show that G is isomorphic to $F(X)$.

We define a mapping λ from G to $F(X)$ as follows. If g is any element of G, then, according to Theorem 5.4, g can be uniquely expressed in the form $g = \sum_{x \in X} a_x x$ where $(a_x)_{x \in X}$ is a quasi-finite family of elements of R. The element $\lambda(g)$ of $F(X)$ is defined by setting for each element t of X,

$$[\lambda(g)]\,(t) = a_t.$$

It is easily verified that λ is an isomorphism. (If we identify X with its image in $F(X)$ under the natural mapping φ and write the elements of $F(X)$ as formal sums then of course $F(X) = G$ and λ is the identity mapping.)

(2) Conversely, suppose G is free. Then there is a set E and an isomorphism μ from $F(E)$ onto G. We shall show that $\mu(E)$ is a basis for G.

Let us write the elements of $F(E)$ as formal sums. Since μ is surjective, if g is any element of G there is an element $\sum_{x \in E} a_x x$ of $F(E)$ such that

$$g = \mu(\sum a_x x) = \sum a_x \mu(x);$$

this shows that $\mu(E)$ is a generating system for G. To show that $\mu(E)$ is linearly independent, suppose $(b_x)_{x \in E}$ is a quasi-finite family of elements of R such that $\sum b_x \mu(x) = 0$. Then $\mu(\sum b_x x) = 0$ and hence, since μ is injective, $\sum b_x x = 0$ in $F(E)$, i.e. all the elements b_x are zero.

Thus $\mu(E)$ is indeed a basis for G.

COROLLARY 1. *For every ring R the left R-module R_l is free.*

Proof. The set consisting of the identity alone is a basis.

COROLLARY 2. *If R is a division ring every left R-module (left vector space over R) is free.*

Proof. The Corollary to Theorem 5.5 shows that every such module has a basis.

In the next two theorems we establish two basic properties of free modules.

THEOREM 9.3. *If V is any left R-module there exists an epimorphism from a free R-module onto V.*

Proof. Let $F = F(V)$ be the free left R-module based on V, φ the natural mapping from V to $F(V)$.

Since (F, φ) is universal for mappings from V to left R-modules (Theorem 9.1) and the identity mapping I_V is a mapping from V to the left R-module V there exists a homomorphism v from F to V such that $v\varphi = I_V$. Then, if x is any element of V, we have $x = I_V(x) = v(\varphi(x))$; so v is an epimorphism.

The construction we have just given is by no means unique, and in fact it is an extremely uneconomical one. The reader may care to verify that if X is any generating system for V then V is an epimorphic image of the free module $F(X)$—instead of the identity mapping I_V in the argument above we use the canonical injection of X into V.

THEOREM 9.4. *Let V and V'' be left R-modules, η an epimorphism from V onto V'', F a free left R-module and α a homomorphism from F to V''. Then there exists a homomorphism β from F to V such that $\eta\beta = \alpha$.*

Proof. Let X be a basis for F. Since η is an epimorphism it follows that for every element x of X there is an element v_x of V such that $\eta(v_x) = \alpha(x)$. We now define a mapping β from F to V as follows. If t is any element of F it may be expressed uniquely in the form $t = \sum_{x \in X} a_x x$ where $(a_x)_{x \in X}$ is as usual a quasi-finite family of elements of R. Set $\beta(t) = \sum_{x \in X} a_x v_x$. Then it is easy to verify that β is a homomorphism and $\eta\beta = \alpha$.

The property established for free modules in Theorem 9.4 is of such importance that we give a special name to those modules which enjoy it. In introducing this we restate the property in the more transparent language of diagrams: a left R-module P is said to be *projective* if in every diagram

$$
\begin{array}{c}
P \\
\downarrow \alpha \\
V \xrightarrow{\ \eta\ } V'' \longrightarrow 0
\end{array}
$$

of modules and homomorphisms in which the row is exact we can insert a homomorphism $\beta : P \to V$ such that the diagram

$$
\begin{array}{c}
\quad P \\
\beta \swarrow \downarrow \alpha \\
V \xrightarrow{\ \eta\ } V'' \longrightarrow 0
\end{array}
$$

is commutative. (We say in this situation that α has been *lifted* to β.) We may thus restate Theorem 9.4 briefly in the following way. *Every free left R-module is projective.* Theorem 9.3 then allows us to say that *every left R-module is an epimorphic image of a projective module* (though Theorem 9.3 is of course a stronger result than this, this is what we shall use later). Our next theorem gives us two useful criteria for a left R-module to be projective.

THEOREM 9.5. *Let P be a left R-module; then the following conditions are equivalent:*

(a) *P is projective;*

(b) *every short exact sequence of left R-modules of the form*

$$0 \to V' \to V \to P \to 0$$

splits;

(c) *P is a direct summand of a free left R-module.*

Proof. (1) Suppose P is a projective left R-module. Let

$$0 \to V' \to V \xrightarrow{\eta} P \to 0 \qquad [9.3]$$

be a short exact sequence, so that η is an epimorphism. Consider the identity mapping I_P from P to P. Since P is projective this mapping can be lifted to a homomorphism η' from P to V, i.e. there is a homomorphism η' from P to V such that $\eta\eta' = I_P$. According to Theorem 8·6, the sequence [9.3] splits (η' is clearly injective).

Thus (a) implies (b).

(2) Suppose condition (b) is satisfied.

In the proof of Theorem 9.3 we showed how to construct a short exact sequence

$$0 \to \operatorname{Ker} v \to F(P) \xrightarrow{v} P \to 0.$$

According to condition (b) this sequence splits and hence, as we remarked after proving Theorem 8.6, the direct sum $\operatorname{Ker} v \oplus P$ is isomorphic to the free module $F(P)$—and hence is free. So P is a direct summand of a free left R-module.

Thus (b) implies (c).

(3) Finally suppose condition (c) is satisfied. Then there exists a free left R-module F and a submodule P' of F such that F is the internal direct sum of P and P'. It follows that the external direct sum $S = P \oplus P'$ is free, and hence projective. Let π be the canonical projection epimorphism from S onto P, ι the canonical injection monomorphism from P to S; we recall that $\pi\iota = I_P$.

Now let V and V'' be any two modules, η an epimorphism from V onto V'', α a homomorphism from P to V''. Then $\alpha\pi$ is a homomorphism from S to V'', and since S is projective there is therefore a homomorphism β_1 from S to V such that $\eta\beta_1 = \alpha\pi$. Set $\beta = \beta_1\iota$;

then β is a homomorphism from P to V and we have $\eta\beta = \eta\beta_1 \iota = \alpha\pi\iota = \alpha I_P = \alpha$. So P is projective.

Hence (c) implies (a) and the proof is complete.

We return to one of the questions raised at the end of §7 which we are now in a position to answer.

THEOREM 9.6. *Let M be a left R-module and let*

$$0 \to V' \xrightarrow{\alpha} V \xrightarrow{\beta} V'' \to 0 \qquad [9.4]$$

be a short exact sequence of left R-modules. If M is projective then the sequence

$$0 \to \mathrm{Hom}\,(M, V') \xrightarrow{\alpha_*} \mathrm{Hom}\,(M, V) \xrightarrow{\beta_*} \mathrm{Hom}\,(M, V'') \to 0 \quad [9.5]$$

is exact. Conversely, if [9.5] *is exact for every short exact sequence* [9.4], *then M is projective.*

Proof. It follows from Theorem 7.4 that the sequence [9.5] is exact if and only if β_* is an epimorphism.

(1) Suppose M is projective and let θ'' be any element of $\mathrm{Hom}\,(M, V'')$. Then we have a diagram

$$\begin{array}{c} M \\ \downarrow{\scriptstyle\theta''} \\ V \xrightarrow{\beta} V'' \to 0 \end{array} \qquad [9.6]$$

in which the row is exact. Since M is projective, there exists a homomorphism θ from M to V such that $\beta\theta = \theta''$. But, by definition, $\beta_*(\theta) = \beta\theta$, so we have indeed shown that $\theta'' \in \mathrm{Im}\,\beta_*$, i.e. that β_* is an epimorphism and so [9.5] is exact.

(2) Conversely, suppose that for every short exact sequence [9.4] the induced homomorphism β_* is an epimorphism. Let [9.6] be a diagram in which the row is exact. Since β_* is an epimorphism from $\mathrm{Hom}\,(M, V)$ onto $\mathrm{Hom}\,(M, V'')$ and $\theta'' \in \mathrm{Hom}\,(M, V'')$ it follows that there is a homomorphism θ from M to V such that $\beta_*(\theta) = \theta''$, i.e. such that $\beta\theta = \theta''$. Thus M is projective.

We next introduce a property of modules which, in a technical sense* is dual to the projective property. A left R-module Q is said

* See Adamson, *Rings, modules and algebras*, §15.

to be *injective* if in every diagram

$$0 \longrightarrow V' \overset{\kappa}{\longrightarrow} V$$
$$\alpha \downarrow$$
$$Q$$

of modules and homomorphisms in which the row is exact we can insert a homomorphism $\beta: V \to Q$ such that the diagram

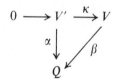

is commutative; β is called an *extension* of α to V.

Example 1. Let D be a division ring. We shall show that every left D-module (left vector space over D) is injective. So let Q be any left D-module and suppose we have a diagram

$$0 \longrightarrow V' \overset{\kappa}{\longrightarrow} V$$
$$\alpha \downarrow$$
$$Q$$

with exact row, i.e. a monomorphism κ from V' to V and a homomorphism α from V' to Q. According to the Corollary of Theorem 5.5 there exists a basis B' for V'; since κ is a monomorphism, $\kappa(B')$ is a free subset of V. Theorem 5.5 itself now shows that there exists a subset B'' of V, disjoint from $\kappa(B')$, such that $\kappa(B') \cup B''$ is a basis for V; then every element x of V can be expressed uniquely in the form

$$x = \sum_{s \in B'} a_s \kappa(s) + \sum_{t \in B''} a_t t,$$

where $(a_s)_{s \in B'}$ and $(a_t)_{t \in B''}$ are quasi-finite families of elements of R.

Define a mapping β from V to Q by setting

$$\beta(x) = \sum_{s \in B'} a_s \alpha(s)$$

for each element x of V. Then it is clear that β is a homomorphism and that $\beta \kappa = \alpha$. So Q is injective, as we claimed.

We shall obtain another example once we have established the following criterion for a module to be injective.

THEOREM 9.7. *Let Q be a left R-module. Then Q is injective if and only if for every left ideal L of R (considered as left R-module) and every homomorphism α from L to Q there exists an element a of Q such that $\alpha(l) = la$ for all elements l of L.*

Proof. (1) Suppose Q is injective.
Consider the diagram

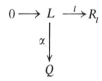

in which L is a left ideal of R, ι is the inclusion monomorphism *inj* from L to R and α is a homomorphism from L to Q. Since, by hypothesis, Q is injective, there exists a homomorphism β from R_l to Q such that $\beta \iota = \alpha$. Then for each element l of L we have $\alpha(l) = \beta \iota(l) = l\beta \iota(e)$; so $\beta \iota(e)$ is the required element a of Q.

(2) Conversely, suppose Q satisfies the condition of the theorem.

Let κ be a monomorphism from a left R-module V' to another left R-module V and let α be a homomorphism from V' to Q. Making use of Zorn's Lemma we shall show that there exists a homomorphism β from V to Q such that $\beta \kappa = \alpha$.

We consider the set E of all ordered pairs (W, φ) consisting of a submodule W of V which includes $\kappa(V')$ and a homomorphism φ from W to Q such that $\varphi \kappa = \alpha$. Write $(W, \varphi) \leqslant (W', \varphi')$ if W is a submodule of W' and $\varphi'(w) = \varphi(w)$ for every element w of W. The reader is invited to verify that E is inductively ordered by the relation $\leqslant$; according to Zorn's Lemma, E has a maximal element, say (V_0, φ_0). We claim that $V_0 = V$; once we have established this,

D

φ_0 will serve as the required homomorphism β. Suppose, to the contrary, that $V_0 \neq V$ and let x be an element of V which does not belong to V_0. If V_1 is the submodule of V generated by V_0 and x, every element y of V_1 can be expressed in the form

$$y = y_0 + rx, \qquad [9.7]$$

where $y_0 \in V_0$ and $r \in R$. We propose to define a homomorphism φ_1 from V_1 to Q such that $\varphi_1(y_0) = \varphi_0(y_0)$ for all elements y_0 of V_0; we shall then have $(V_1, \varphi_1) > (V_0, \varphi_0)$, contradicting the maximal property of (V_0, φ_0).

To this end, let L be the set of elements r of the ring R such that $rx \in V_0$; we see at once that L is a left ideal of R. Consider the mapping α from L to Q defined by setting $\alpha(l) = \varphi_0(lx)$ for all elements l of L. This mapping is clearly a homomorphism and hence, by hypothesis, there is an element a of Q such that $\varphi_0(lx) = \alpha(l) = la$ for all elements l of L. We now define a mapping φ_1 from V_1 to Q by setting, for each element $y = y_0 + rx$ of V_1,

$$\varphi_1(y) = \varphi_0(y_0) + ra.$$

This is independent of the expression [9.7] for y; if we have $y = y_0 + rx = y_0' + r'x$, then $(r - r') x = y_0' - y_0 \in V_0$ and hence

$$\varphi_0(y_0') - \varphi_0(y_0) = \varphi_0(y_0' - y_0) = \varphi_0((r - r') x) = (r - r') a$$
$$= ra - r'a,$$

so that

$$\varphi_0(y_0) + ra = \varphi_0(y_0') + r'a.$$

It is now a trivial matter to check that φ_1 is a homomorphism satisfying the required property. Thus our contradiction is established.

Example 2. An additive abelian group A is said to be *divisible* if for every element x of A and every non-zero integer m there exists an element x' of A such that $x = mx'$. We shall show that a **Z**-module A is injective if and only if it is divisible. This will follow from Theorem 9.7 if we can show that the property of divisibility is equivalent to that described in the theorem.

First, suppose that A is divisible and let L be an ideal of **Z**, α a homomorphism from L to A. Since all the ideals of **Z** are principal there is an integer m such that every element l of L has the form $l = km$ where $k \in \mathbf{Z}$. Let $\alpha(m) = b$; since A is divisible, there is an element a of A such that $b = ma$. (If $m = 0$ we have $b = 0$ and we take $a = 0$.) Now we have, for each element l of L, $\alpha(l) = \alpha(km) = k\alpha(m) = kb = kma = la$. Thus A satisfies the condition of Theorem 9.7 and hence is injective.

Conversely, suppose that A is injective. Let b be any element of A, m any non-zero integer. If L is the principal ideal of $\mathbf{Z}$ generated by m, we may define a homomorphism α from L to A by setting $\alpha(km) = kb$ for each element km of L. According to Theorem 9.7 there is an element a of A such that $\alpha(l) = la$ for every element l of L. We have in particular $b = \alpha(m) = ma$; so A is divisible.

We can now deal with the second of the questions raised at the end of §7.

THEOREM 9.8. *Let M be a left R-module and let*

$$0 \to V' \xrightarrow{\alpha} V \xrightarrow{\beta} V'' \to 0 \qquad [9.8]$$

be a short exact sequence of left R-modules. If M is injective then the sequence

$$0 \to \operatorname{Hom}(V'', M) \xrightarrow{\beta^*} \operatorname{Hom}(V, M) \xrightarrow{\alpha^*} \operatorname{Hom}(V', M) \to 0 \qquad [9.9]$$

is exact. Conversely, if [9.9] is exact for every short exact sequence [9.8] then M is injective.

Proof. It follows from Theorem 7.4 that the sequence [9.9] is exact if and only if α^* is an epimorphism.

(1) Suppose M is injective and let θ' be any element of $\operatorname{Hom}(V', M)$. Then we have a diagram

$$\begin{array}{ccc} 0 \longrightarrow & V' & \xrightarrow{\alpha} V \\ & \theta' \downarrow & \\ & M & \end{array} \qquad [9.10]$$

in which the row is exact. According to the injective property of M there exists a homomorphism θ from V to M such that $\theta\alpha = \theta'$. But, by definition of α^*, we have $\alpha^*(\theta) = \theta\alpha$; so $\theta' = \alpha^*(\theta)$. Hence α^* is an epimorphism and [9.9] is exact.

(2) Conversely, suppose that for every short exact sequence [9.8] the induced homomorphism α^* from $\operatorname{Hom}(V, M)$ to $\operatorname{Hom}(V', M)$, is an epimorphism. Let [9.10] be a diagram in which the row is exact. Then, since α^* is an epimorphism and $\theta' \in \operatorname{Hom}(V', M)$,

there is a homomorphism θ in Hom (V, M) such that $\alpha^*(\theta) = \theta'$, i.e. such that $\theta\alpha = \theta'$. Hence M is injective.

We proved earlier that for every left R-module V there exists a projective module P and an epimorphism from P onto V. We state, without proof, the corresponding ('dual') result for injective modules.

THEOREM 9.9. *Let V be any left R-module. Then there exist an injective module Q and a monomorphism from V to Q.*

If we were to identify V with its image under the monomorphism we could say that every left R-module is a submodule of an injective module.

As a consequence of this theorem we obtain another criterion for a module to be injective, analogous to one of those in Theorem 9.5 for projective modules.

THEOREM 9.10. *A left R-module Q is injective if and only if every short exact sequence of the form*

$$0 \to Q \to V \to V'' \to 0$$

splits.

Proof. (1) Suppose Q is injective. Let

$$0 \to Q \xrightarrow{\kappa} V \to V'' \to 0 \qquad [9.11]$$

be a short exact sequence, so that κ is a monomorphism. Consider the identity mapping I_Q from Q onto Q. Since Q is injective there is an extension κ' of this mapping to V, i.e. a homomorphism κ' from V to Q such that $\kappa'\kappa = I_Q$. Then, according to Theorem 8.6 the sequence [9.11] splits.

(2) Conversely, suppose every such sequence splits.

According to Theorem 9.9 there exists an injective module Q_1 and a monomorphism κ from Q to Q_1. Then the sequence

$$0 \to Q \xrightarrow{\kappa} Q_1 \xrightarrow{\eta} Q_1/\kappa(Q) \to 0$$

(where η is the canonical epimorphism from Q_1 onto $Q_1/\kappa(Q)$) is exact and therefore splits. Hence there is a homomorphism κ' from Q_1 to Q such that $\kappa'\kappa = I_Q$.

Now suppose we have a diagram

$$0 \longrightarrow V' \stackrel{\lambda}{\longrightarrow} V$$

$$\alpha \Big\downarrow$$

$$Q$$

in which the row is exact. Then $\kappa\alpha$ is a homomorphism from V' to the injective module Q_1 and so there exists a homomorphism β from V to Q_1 such that $\beta\lambda = \kappa\alpha$. It follows that $(\kappa'\beta) \lambda = (\kappa'\kappa) \alpha = I_Q\alpha = \alpha$. Thus Q is injective.

§10. Tensor Products

Let R be a ring with identity, V a right R-module and W a left R-module. We form the free $\mathbf{Z}$-module $F = F(V \times W)$ based on the Cartesian product of V and W; as in §9, we shall write the elements of F as 'formal sums' $\sum_{(x,y)\in V \times W} n_{(x,y)}(x, y)$, where the family of integers $(n_{(x,y)})$ is quasi-finite, i.e., only finitely many of the integers $n_{(x,y)}$ are non-zero. In particular, if $n_{(x_1,y_1)} = 1$ and $n_{(x,y)} = 0$ for $(x, y) \neq (x_1, y_1)$ we shall abbreviate the formal sum simply to (x_1, y_1).

Consider the subgroup K of F generated by the set of all elements of the forms

$$\begin{aligned}
(x_1 + x_2, y) &- (x_1, y) - (x_2, y), \\
(x, y_1 + y_2) &- (x, y_1) - (x, y_2), \qquad [10.1] \\
(xa, y) &- (x, ay),
\end{aligned}$$

where $x, x_1, x_2 \in V$, $y, y_1, y_2 \in W$ and $a \in R$. The factor group F/K is called the *tensor product* of V and W and is denoted by $V \otimes_R W$, or simply by $V \otimes W$ if the ring R is clear from the context. We read $V \otimes_R W$ as 'V-tensor-R-W'. Comparison between this construction and others (such as the homomorphism groups discussed in §7) would be facilitated by the introduction of the notation $\mathrm{Ten}_R(V, W)$ instead of $V \otimes_R W$; but the latter is too well established for there to be any hope of changing it.

Let η be the canonical epimorphism from F onto F/K. For each ordered pair of elements (x, y) in $V \times W$ we denote $\eta(x, y)$ by $x \otimes y$ (read 'x-tensor-y'). Then every element of the tensor product has the form

$$\eta\left(\sum n_{(x, y)}(x, y)\right) = \sum n_{(x, y)} x \otimes y$$

where $(n_{(x, y)})$ is a quasi-finite family of integers; clearly we may write this alternatively in the form

$$\sum_{i \in I} x_i \otimes y_i$$

where (x_i) and (y_i) are families of elements of V and W respectively indexed by the same finite index set I.

Since all elements of the forms [10.1] belong to the kernel of η it follows that we have

$$\begin{aligned}
(x_1 + x_2) \otimes y &= x_1 \otimes y + x_2 \otimes y, \\
x \otimes (y_1 + y_2) &= x \otimes y_1 + x \otimes y_2, \qquad [10.2] \\
xa \otimes y &= x \otimes ay,
\end{aligned}$$

for all elements x, x_1, x_2 of V, y, y_1, y_2 of W and a of R. We deduce that for every element y of W we have $0 \otimes y = 0$; namely

$$0 \otimes y = (0 + 0) \otimes y = 0 \otimes y + 0 \otimes y$$

and hence $0 \otimes y = 0$ as asserted. Similarly we have $x \otimes 0 = 0$ for every element x of V.

Still let V and W be right and left R-modules respectively, and let A be any additive abelian group. A mapping φ from the Cartesian product $V \times W$ to A is said to be *balanced* if for all elements x, x_1, x_2 of V, y, y_1, y_2 of W and a of R we have

$$\begin{aligned}
\varphi(x_1 + x_2, y) &= \varphi(x_1, y) + \varphi(x_2, y), \\
\varphi(x, y_1 + y_2) &= \varphi(x, y_1) + \varphi(x, y_2), \\
\varphi(xa, y) &= \varphi(x, ay).
\end{aligned}$$

For example, the equations [10.2] show that the mapping v from $V \times W$ to $V \otimes W$ defined by setting $v(x, y) = x \otimes y$ for every ordered pair (x, y) in $V \times W$ is a balanced mapping; we call it the *natural mapping* from $V \times W$ to $V \otimes W$.

À pair (B, β) consisting of an additive abelian group B and a balanced mapping β from $V \times W$ to B is said to be *universal* for balanced mappings from $V \times W$ if for every balanced mapping φ from $V \times W$ to an additive abelian group A there exists a unique abelian group homomorphism α from B to A such that the diagram

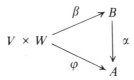

is commutative.

THEOREM 10.1. *Let V and W be right and left R-modules respectively. Then the pair $(V \otimes W, v)$ consisting of the tensor product of V and W and the natural mapping from $V \times W$ to $V \otimes W$ is universal for balanced mappings from $V \times W$.*

Proof. Let φ be a balanced mapping from $V \times W$ to an additive abelian group A. According to Theorem 9.1 there is a unique $\mathbf{Z}$-module (abelian group) homomorphism φ' from the free $\mathbf{Z}$-module $F(V \times W)$ to A such that $\varphi'(x, y) = \varphi(x, y)$ for every pair (x, y) in $V \times W$. (As usual we are identifying each pair in $V \times W$ with its image under the natural mapping in $F(V \times W)$.) From the hypothesis that φ is balanced, it follows easily that all the elements of K are contained in the kernel of the homomorphism φ': for instance, we have

$$\varphi'((x_1 + x_2, y) - (x_1, y) - (x_2, y))$$
$$= \varphi'(x_1 + x_2, y) - \varphi'(x_1, y) - \varphi'(x_2, y)$$
$$= \varphi(x_1 + x_2, y) - \varphi(x_1, y) - \varphi(x_2, y) = 0.$$

We now define a mapping α from $V \otimes W$ to A as follows. If C is any element of $V \otimes W$, i.e. a coset of $F(V \times W)$ modulo K, choose any element c from C and set $\alpha(C) = \varphi'(c)$; this clearly depends only on C, not on the choice of the element c, for if c' is another element in C we have $c - c' \in K \subseteq \operatorname{Ker} \varphi'$ and so $\varphi'(c) = \varphi'(c')$. This mapping α is easily seen to be a homomorphism, and we clearly have $\alpha v = \varphi$.

To see that α is the unique homomorphism satisfying this condition, let α' be another homomorphism from $V \otimes W$ to A such that $\alpha'v = \varphi$. Let C be any element of $V \otimes W$; then, as we have seen, C

can be expressed in the form $C = \sum_{i \in I} x_i \otimes y_i = \sum_{i \in I} v(x_i, y_i)$. Then we have

$$\alpha(C) = \alpha\left(\sum v(x_i, y_i)\right) = \sum \alpha v(x_i, y_i) = \sum \varphi(x_i, y_i)$$

and similarly

$$\alpha'(C) = \sum \varphi(x_i, y_i).$$

Thus $\alpha = \alpha'$, as asserted.

Suppose for a moment that R is a commutative ring, and let V and W be right and left R-modules respectively. Let r be any element of R and consider the mapping φ_r from $V \times W$ to $V \otimes W$ defined by setting $\varphi_r(x, y) = xr \otimes y = x \otimes ry$ for each ordered pair (x, y) in $V \times W$; we contend that φ_r is balanced. So let x, x_1, x_2 be elements of V, y, y_1, y_2 elements of W and a an element of R. Then

$$\varphi_r(x_1 + x_2, y) = ((x_1 + x_2)r) \otimes y = (x_1 r + x_2 r) \otimes y$$
$$= (x_1 r) \otimes y + (x_2 r) \otimes y$$
$$= \varphi_r(x_1, y) + \varphi_r(x_2, y);$$

similarly

$$\varphi_r(x, y_1 + y_2) = \varphi_r(x, y_1) + \varphi_r(x, y_2);$$

further

$$\varphi_r(xa, y) = (xa)r \otimes y = x(ar) \otimes y \stackrel{*}{=} x(ra) \otimes y = (xr)a \otimes y$$
$$= xr \otimes ay = \varphi_r(x, ay).$$

The commutativity of R is used at the step marked with an asterisk. It follows that there is a homomorphism α_r from $V \otimes W$ to $V \otimes W$ such that $\alpha_r(x \otimes y) = \varphi_r(x, y) = xr \otimes y$ for every ordered pair (x, y) in $V \times W$.

We now define a scalar multiplication on the left of $V \otimes W$ by elements of R by setting, for each element t of $V \otimes W$ and each element r of R,

$$rt = \alpha_r(t)$$

and claim that under this scalar multiplication the abelian group $V \otimes W$ is turned into a left R-module.

To establish this result, we remark that since α_r is a homomorphism we have

$$r(t_1 + t_2) = \alpha_r(t_1 + t_2) = \alpha_r(t_1) + \alpha_r(t_2) = rt_1 + rt_2$$

for every pair of elements t_1, t_2 of $V \otimes W$. It now remains to show that for every element t of $V \otimes W$ and every pair of elements r_1, r_2 of R we have

$$(r_1 + r_2) t = r_1 t + r_2 t, \quad (r_1 r_2) t = r_1(r_2 t), \quad et = t.$$

Since, as we have seen, every element t of $V \otimes W$ can be expressed as the sum of a finite family of elements of the form $x \otimes y$, it is sufficient to establish these relations for elements of this form— and this is trivial.

If R is once more a general ring (i.e. not necessarily commutative) then tensor products $V \otimes W$ are in general only abelian groups, not R-modules. But there are two special cases to consider even in the general situation, where tensor products can be given an R-module structure. These are the cases $V = R_r$ (R itself, considered as a right R-module) and $W = R_l$ (R as left R-module). We consider the first of these in some detail. Let b be any element of R and consider the mapping φ_b from $R \times W$ to $R_r \otimes W$ defined by setting $\varphi_b(a, y) = ba \otimes y$ for each ordered pair (a, y) in $R \times W$. Routine computations show that φ_b is a balanced mapping; hence there is a homomorphism α_b from $R_r \otimes W$ to itself such that $\alpha_b v = \varphi_b$, i.e. such that $\alpha_b(a \otimes y) = ba \otimes y$. We define a scalar multiplication on the left of $R_r \otimes W$ by elements of R by setting $bt = \alpha_b(t)$ for each element b of R and each element t of $R_r \otimes W$; under this scalar multiplication $R_r \otimes W$ becomes a left R-module. By a similar device we turn $V \otimes R_l$ into a right R-module.

THEOREM 10.2. *Let R be a ring with identity, V and W right and left R-modules respectively. Then the left R-module $R_r \otimes W$ is isomorphic to W and the right R-module $V \otimes R_l$ is isomorphic to V.*

Proof. Consider the mapping φ from $R \times W$ to W defined by setting $\varphi(a, y) = ay$ for each ordered pair (a, y) in $R \times W$. Since W is a left R-module this mapping φ is balanced, and hence there exists an abelian group homomorphism α from $R_r \otimes W$ to W such that $\alpha v = \varphi$. We claim that when $R_r \otimes W$ is considered as a left R-module in the manner described above this mapping α is actually an R-module homomorphism. To see this, first let a, b be elements of R, y an element of W; then we have

$$\alpha(b(a \otimes y)) = \alpha((ba) \otimes y) = \alpha(v(ba, y)) = \varphi(ba, y) = (ba) y$$

while

$$b\alpha(a \otimes y) = b\alpha(v(a, y)) = b\varphi(a, y) = b(ay) = (ba)\,y.$$

So $\alpha(b(a \otimes y)) = b\alpha(a \otimes y)$. Then, since every element t of $R_r \otimes W$ can be expressed as a sum of a finite family of elements of the form $a \otimes y$, it follows easily than $\alpha(bt) = b\alpha(t)$ for all elements b of R and all elements t of $R_r \otimes W$. Thus α is indeed an R-module homomorphism.

To show that α is an epimorphism, let y be any element of W. Then $y = ey = \varphi(e, y) = \alpha(v(e, y)) = \alpha(e \otimes y)$.

Next we show that α is a monomorphism. Let t be any element of Ker α. Then t can be expressed in the form $t = \sum_{i=1}^{n} a_i \otimes y_i$ $= \sum_{i=1}^{n} ea_i \otimes y_i = \sum_{i=1}^{n} e \otimes a_i y_i = e \otimes \sum_{i=1}^{n} a_i y_i = e \otimes y$ say, and we have $0 = \alpha(t) = ey = y$. Thus $t = e \otimes 0 = 0$, as required.

Hence α is an isomorphism from the left R-module $R_r \otimes W$ onto W. Similarly we set up an isomorphism from the right R-module $V \otimes R_l$ onto V.

Let V_1 and V_2 be right R-modules, W_1 and W_2 left R-modules, α and β homomorphisms from V_1 to V_2 and W_1 to W_2 respectively. Consider the mapping φ from $V_1 \times W_1$ to $V_2 \otimes W_2$ defined by setting $\varphi(x_1, y_1) = \alpha(x_1) \otimes \beta(y_1)$ for every ordered pair (x_1, y_1) in $V_1 \times W_1$. We easily verify that φ is a balanced mapping; hence there exists an abelian group homomorphism γ from $V_1 \otimes W_1$ to $V_2 \otimes W_2$ such that $\gamma v = \varphi$, i.e. such that $\gamma(x_1 \otimes y_1) = \alpha(x_1) \otimes \beta(y_1)$ for every ordered pair (x_1, y_1) in $V_1 \times W_1$. We denote this homomorphism γ by $\alpha \otimes \beta$ and call it the *tensor product* of α and β.

In the next two theorems we establish fundamental properties of the tensor product of homomorphisms.

THEOREM 10.3. *Let V_1, V_2, V_3 be right R-modules, W_1, W_2, W_3 be left R-modules and let α, α', β, β' be homomorphisms from V_1 to V_2, V_2 to V_3, W_1 to W_2 and W_2 to W_3 respectively. Then $(\alpha' \otimes \beta')(\alpha \otimes \beta) = (\alpha'\alpha) \otimes (\beta'\beta)$.*

Proof. This follows at once from the consideration that for every ordered pair (x_1, y_1) in $V_1 \times W_1$ we have

$$(\alpha' \otimes \beta') [(\alpha \otimes \beta)(x_1 \otimes y_1)] = (\alpha' \otimes \beta')(\alpha(x_1) \otimes \beta(y_1))$$
$$= \alpha'(\alpha(x_1)) \otimes \beta'(\beta(y_1))$$
$$= \alpha'\alpha(x_1) \otimes \beta'\beta(y_1).$$

THEOREM 10.4. *Let V_1, V_2 be right R-modules, W_1, W_2 left R-modules; let α, α_1, α_2 be homomorphisms from V_1 to V_2 and β, β_1, β_2 homomorphisms from W_1 to W_2. Then*

$$(\alpha_1 + \alpha_2) \otimes \beta = \alpha_1 \otimes \beta + \alpha_2 \otimes \beta$$

and

$$\alpha \otimes (\beta_1 + \beta_2) = \alpha \otimes \beta_1 + \alpha \otimes \beta_2.$$

Proof. For every ordered pair (x_1, y_1) in $V_1 \times W_1$ we have

$$[(\alpha_1 + \alpha_2) \otimes \beta] (x_1 \otimes y_1) = (\alpha_1 + \alpha_2)(x_1) \otimes \beta(y_1)$$
$$= (\alpha_1(x_1) + \alpha_2(x_1)) \otimes \beta(y_1)$$
$$= \alpha_1(x_1) \otimes \beta(y_1) + \alpha_2(x_1) \otimes \beta(y_1)$$
$$= (\alpha_1 \otimes \beta)(x_1 \otimes y_1) + (\alpha_2 \otimes \beta)(x_1 \otimes y_1)$$
$$= [\alpha_1 \otimes \beta + \alpha_2 \otimes \beta] (x_1 \otimes y_1).$$

Thus $(\alpha_1 + \alpha_2) \otimes \beta = (\alpha_1 \otimes \beta) + (\alpha_2 \otimes \beta)$ and the other result follows similarly.

We now examine the effect on a short exact sequence of right R-modules of forming the tensor products of its terms with a fixed left R-module.

THEOREM 10.5. *Let*

$$0 \rightarrow V' \xrightarrow{\alpha} V \xrightarrow{\beta} V'' \rightarrow 0$$

be an exact sequence of right R-modules; let M be a left R-module. Then the sequence

$$V' \otimes M \xrightarrow{\alpha \otimes I_M} V \otimes M \xrightarrow{\beta \otimes I_M} V'' \otimes M \rightarrow 0$$

is exact.

Proof. Let us abbreviate $\alpha \otimes I_M$ and $\beta \otimes I_M$ to α_* and β_* respectively. We must show that (a) Im $\alpha_* \subseteq$ Ker β_*, (b) Ker $\beta_* \subseteq$ Im α_*, and (c) β_* is an epimorphism.

(a) Let t be any element of Im α_*; then there is an element t' of $V' \otimes M$ such that $t = \alpha_*(t')$. This element t' can be expressed in the form $t' = \sum_{i \in I} x_i' \otimes m_i$ where $(x_i')_{i \in I}$, $(m_i)_{i \in I}$ are finite families of elements of V' and M respectively. For each index i in I we have

$$\beta_*(\alpha_*(x_i' \otimes m_i)) = \beta_*(\alpha(x_i') \otimes m_i) = \beta(\alpha(x_i')) \otimes m_i = 0 \otimes m_i = 0.$$

It follows that $\beta_*(t) = \beta_*(\alpha_*(t')) = 0$ and so Im $\alpha_* \subseteq \text{Ker } \beta_*$ as required.

(b) To show that Ker $\beta_* \subseteq$ Im α_* we begin by defining an abelian group homomorphism γ from Coker $\alpha_* = (V \otimes M)/\text{Im } \alpha_*$ to $V'' \otimes M$ as follows. If C is any coset of $V \otimes M$ modulo Im α_*, choose an element t from C and define $\gamma(C)$ to be $\beta_*(t)$; then $\gamma(C)$ is clearly independent of the choice of t, for if t_1 is another element in C we have $t - t_1 \in \text{Im } \alpha_* \subseteq \text{Ker } \beta_*$ and hence $\beta_*(t_1) = \beta_*(t)$.

Next consider the mapping φ from $V'' \times M$ to Coker α_* defined by setting $\varphi(t'', m) = $ coset of $t \otimes m$ modulo Im α_* where t is any element of $\beta^{-1}(t'')$. Then $\varphi(t'', m)$ so defined is independent of the choice of t; for if t_1 is another element of $\beta^{-1}(t'')$ we have $t - t_1 \in \text{Ker } \beta = \text{Im } \alpha$ and hence $(t - t_1) \otimes m \in \text{Im } \alpha_*$. It is easy to verify that φ is a balanced mapping. So there exists a homomorphism γ' from $V'' \otimes M$ to Coker α_* such that $\gamma'(t'' \otimes m) = \varphi(t'', m)$ for every pair (t'', m) in $V'' \times M$.

If C is any element of Coker α_* we clearly have $\gamma'\gamma(C) = C$. So γ is a monomorphism. But it follows at once from the definition that Ker $\gamma = \text{Ker } \beta_*/\text{Im } \alpha_*$. Hence Ker $\beta_* = \text{Im } \alpha_*$ as required.

(c) Let t'' be any element of $V'' \otimes M$; t'' can be expressed in the form $t'' = \sum_{i \in I} x_i'' \otimes m_i$ where (x_i'') and (m_i) are finite families of elements of V'' and M respectively. Since β is an epimorphism it follows that for each index i in I there is an element x_i of V such that $x_i'' = \beta(x_i)$. Then we have

$$t'' = \sum \beta(x_i) \otimes m_i = \sum \beta_*(x_i \otimes m_i) = \beta_*(\sum x_i \otimes m_i).$$

Thus β_* is an epimorphism, as required.

This completes the proof.

By an exactly similar argument we can prove the analogue of Theorem 10.5 for short exact sequences of left R-modules.

THEOREM 10 6. *Let*

$$0 \to V' \xrightarrow{\alpha} V \xrightarrow{\beta} V'' \to 0$$

be an exact sequence of left R-modules, M a right R-module. Then the sequence

$$M \otimes V' \xrightarrow{I_M \otimes \alpha} M \otimes V \xrightarrow{I_M \otimes \beta} M \otimes V'' \to 0$$

is exact.

To show that the results of Theorems 10.5 and 10.6 cannot be improved in general we give an example of a right R-module monomorphism α from V' to V say and a left R-module M such that the homomorphism $\alpha \otimes I_M$ from $V' \otimes M$ to $V \otimes M$ is not a monomorphism. Namely, we take $R = \mathbf{Z}$, $V' = \mathbf{Z}_r$, $V = \mathbf{Q}$ (the additive group of rational numbers considered as a $\mathbf{Z}$-module), α the inclusion monomorphism from V' to V, and M an abelian group of order 2; M consists of two elements 0 and m, say, and $2m = 0$. According to Theorem 10.2, the tensor product $V' \otimes M$ is isomorphic to M. But the tensor product $V \otimes M$ is a zero module; for if x is any rational number there exists a rational number x' such that $x = x'2$ (we write it like this since we are thinking of V as a right $\mathbf{Z}$-module); hence we have $x \otimes m = x'2 \otimes m = x' \otimes 2m = x' \otimes 0 = 0$, and of course $x \otimes 0 = 0$. Thus $\alpha \otimes I_M$ is not a monomorphism.

§11. Artinian and Noetherian Modules

Let R be a ring with identity. An infinite sequence $(V_k)_{k \in \mathbf{N}}$ of left R-modules is said to be an *infinite descending chain* if $V_k \supseteq V_{k+1}$ for all natural numbers k; such a chain is said to be *strictly descending* if all the inclusions are proper, i.e. if $V_k \supset V_{k+1}$ for all natural numbers k. A finite sequence $(V_k)_{k \in I}$ of left R-modules, indexed by an interval $I = [p, q]$ of natural numbers is called a *finite descending chain* if $V_k \supseteq V_{k+1}$ for $k = p, \ldots, q - 1$; again the chain is said to be *strictly descending* if all the inclusions are proper. By reversing all the inclusions we obtain definitions for *ascending* and *strictly ascending* chains.

We say that an ascending or descending chain (V_k) *levels off*, or *is eventually constant* if there is an index k_0 such that $V_k = V_{k_0}$ for all indices $k \geq k_0$; we say that the chain levels off at V_{k_0}. Clearly every finite ascending or descending chain levels off at its last term.

A left R-module V is said to satisfy the *descending chain condition* if every descending chain of submodules of V (infinite as well as finite) levels off; clearly V satisfies this condition if and only if there are no infinite strictly descending chains of submodules of V. Similarly V is said to satisfy the *ascending chain condition* if every ascending chain of submodules of V levels off; this is equivalent to requiring that there be no infinite strictly ascending chains of submodules of V.

A left R-module V is said to satisfy the *minimum condition* if every non-empty set of submodules of V has a minimal element with respect to the inclusion relation. Thus V satisfies the minimum condition if and only if in every non-empty set E of submodules of V there is a submodule V_0 such that no submodule in E is properly included in V_0. Similarly V is said to satisfy the *maximum condition* if every non-empty set of submodules of V has a maximal element with respect to the inclusion relation.

THEOREM 11.1. *A left R-module satisfies the minimum condition if and only if it satisfies the descending chain condition; it satisfies the maximum condition if and only if it satisfies the ascending chain condition.*

Proof. We shall prove the first statement of the theorem; the proof of the second part can then be easily obtained by making the obvious modifications.

(1) Suppose V satisfies the minimum condition.

If $(V_k)_{k \in \mathbb{N}}$ were an infinite strictly descending chain of submodules of V the set of all the submodules V_k in this chain would be a non-empty set of submodules of V without a minimal element. Hence there can be no infinite strictly descending chain of submodules of V.

So V satisfies the descending chain condition.

(2) Conversely, suppose V satisfies the descending chain condition.

Let us assume that V does not satisfy the minimum condition and deduce a contradiction; so we suppose that there is a non-empty set E of submodules of V which has no minimal element.

Let V_0 be any element of E. Now suppose that for some natural number k we have constructed a strictly descending chain

$$V_0 \supset V_1 \supset \ldots \supset V_k$$

of submodules in E. Since E has no minimal element there is a submodule V_{k+1} in E such that $V_k \supset V_{k+1}$; so we have a strictly descending chain

$$V_0 \supset V_1 \supset \ldots \supset V_k \supset V_{k+1}$$

of submodules in E. Proceeding in this way, we obtain an infinite strictly descending chain of submodules of V. This, however, contradicts the descending chain condition in V.

Hence V satisfies the minimum condition.

A left R-module which satisfies the minimum condition (and hence the descending chain condition) is called an *Artinian module*; a left R-module which satisfies the maximum condition (and hence the ascending chain condition) is called a *Noetherian module*. These names are given in honour of Emil Artin and Emmy Noether, two of the great pioneers of modern algebra.

We now give an alternative criterion for a module to be Noetherian.

THEOREM 11.2. *A left R-module V is Noetherian if and only if every submodule of V is finitely generated.*

Proof. (1) Suppose every submodule of V is finitely generated.

Let $(V_k)_{k \in N}$ be an infinite ascending chain of submodules of V. Let $W = \bigcup_{k \in N} V_k$. Then W is a submodule of V—for if x and y are elements of W, there are natural numbers n_1, n_2 such that $x \in V_{n_1}$, $y \in V_{n_2}$ and hence $x + y \in V_n \subseteq W$ where $n = \sup(n_1, n_2)$; further, if a is any element of R, we have $ax \in V_{n_1} \subseteq W$.

By hypothesis, W is generated by a finite subset $\{x_1, \ldots, x_r\}$ say. Since the elements $x_1, \ldots, x_r$ belong to the union $\bigcup_{k \in N} V_k$ there exist natural numbers $n_1, \ldots, n_r$ such that $x_i \in V_{n_i}$ ($i = 1, \ldots, r$). Let $k_0 = \sup(n_1, \ldots, n_r)$. Then $\{x_1, \ldots, x_r\} \subseteq V_{k_0}$ and hence $W \subseteq V_{k_0}$; but of course $V_{k_0} \subseteq W$. So $V_{k_0} = W$ and for every natural number $k \geq k_0$ we have $W = V_{k_0} \subseteq V_k \subseteq W$, whence $V_k = V_{k_0}$.

Thus (V_k) levels off. Hence V is Noetherian.

(2) Conversely, suppose V is a Noetherian module.

Let W be any submodule of V and let E be the set of finitely generated submodules of W; E is clearly non-empty since the zero submodule belongs to it. Since V satisfies the maximum condition, E has a maximal element, W_0 say. If we can show that $W_0 = W$ it will follow that W is finitely generated.

Suppose, to the contrary, that $W_0 \neq W$; then there is an element x of W which does not belong to W_0. The submodule W_1 generated by $W_0 \cup \{x\}$ is clearly finitely generated and so belongs to E; but $W_1 \supset W_0$ and this contradicts the maximal property of W_0.

Hence $W = W_0$ and so W is finitely generated, as required.

We show next that the Artinian and Noetherian properties carry over to submodules and factor modules.

THEOREM 11.3. *Let V be an Artinian (Noetherian) left R-module, W a submodule of V. Then W and V/W are Artinian (Noetherian).*

Proof. We shall give the proof in the case where V is an Artinian module; the proof for the Noetherian case is obtained by making the obvious modifications of setting 'maximal' for 'minimal' and 'ascending' for 'descending'.

(1) Let E be a non-empty set of submodules of W. Then of course E is a non-empty set of submodules of V and so, by hypothesis, has a minimal element. Thus W is Artinian.

(2) Let $(\overline{V}_k)_{k \in \mathbb{N}}$ be an infinite descending chain of submodules of V/W. Then $(\eta^{-1}(\overline{V}_k))_{k \in \mathbb{N}}$ is a descending chain of submodules of V (where η is, as usual, the canonical epimorphism from V onto V/W). Since V is Artinian, there is a natural number k_0 such that $\eta^{-1}(\overline{V}_k) = \eta^{-1}(\overline{V}_{k_0})$ for all natural numbers $k \geqslant k_0$. So, since η is an epimorphism, it follows that $\overline{V}_k = \overline{V}_{k_0}$ for all natural numbers $k \geqslant k_0$. Thus V/W is Artinian.

The converse of Theorem 11.3 is true, but the next theorem shows that in order to prove that a module is Artinian (Noetherian) it is unnecessary to assume that all submodules and factor modules are Artinian (Noetherian)—it is enough to have this for one submodule and the corresponding factor module.

THEOREM 11.4. *Let V be a left R-module, W a submodule of V. If W and V/W are Artinian (Noetherian) then V is Artinian (Noetherian).*

Proof. We give the proof this time in the Noetherian case.

Let $(V_k)_{k \in \mathbb{N}}$ be an ascending chain of submodules of V. Then $(W \cap V_k)_{k \in \mathbb{N}}$ and $(\eta(V_k))_{k \in \mathbb{N}}$ are ascending chains of submodules of W and V/W (again η is the canonical epimorphism from V onto V/W). By hypothesis there exist natural numbers k_1, k_2 such that $W \cap V_k = W \cap V_{k_1}$ for all natural numbers $k \geqslant k_1$ and $\eta(V_k) = \eta(V_{k_2})$ for all natural numbers $k \geqslant k_2$. Let $k_0 = \sup(k_1, k_2)$; we claim that $V_k = V_{k_0}$ for all natural numbers $k \geqslant k_0$.

So we let k be any natural number such that $k \geqslant k_0$. We certainly have $V_k \supseteq V_{k_0}$. Suppose next that x is any element of V_k. Then $\eta(x) \in \eta(V_k) = \eta(V_{k_0})$; hence there is an element x_0 of V_{k_0} such that $\eta(x) = \eta(x_0)$, whence $x - x_0 \in W$. But since $V_{k_0} \subseteq V_k$ we also have

$x - x_0 \in V_k$; so $x - x_0 \in W \cap V_k = W \cap V_{k_0}$. It follows that $x \in V_{k_0}$ and so $V_k = V_{k_0}$, as required.

Hence V is Noetherian.

COROLLARY 1. *The external direct sum of a finite family of Artinian (Noetherian) left R-modules is Artinian (Noetherian).*

Proof. Let V_1 and V_2 be Artinian left R-modules, and let $V = V_1 \oplus V_2$. If ι_1 and π_2 have the obvious meanings, we have Ker $\pi_2 = $ Im ι_1 and hence $V/\text{Im } \iota_1 = V/\text{Ker } \pi_2$ is isomorphic to Im $\pi_2 = V_2$. Thus $V/\text{Im } \iota_1$ is Artinian; and Im ι_1, which is isomorphic to V_1, is also Artinian. According to the Theorem it follows that V is Artinian. To obtain the result for finite families with more than two members, we proceed by induction using the easily established fact that $V_1 \oplus \ldots \oplus V_r$ is isomorphic to $(V_1 \oplus \ldots \oplus V_{r-1}) \oplus V_r$ $(r \geqslant 3)$.

COROLLARY 2. *The sum of a finite family of Artinian (Noetherian) submodules of a left R-module is Artinian (Noetherian).*

Proof. The sum of the family of submodules is an epimorphic image of the external direct sum of the family and hence is isomorphic to a factor module of this external direct sum. But the external direct sum is Artinian (Noetherian) by Corollary 1 and so the desired result follows from Theorem 11.3.

The ring R is said to be a *left Artinian ring* if the left R-module R_l is an Artinian module; R is said to be a *left Noetherian ring* if R_l is a Noetherian module. Since the submodules of the left R-module R_l are the left ideals of R (§ 5, Example 8), we see that R is a left Artinian ring if and only if every non-empty set of left ideals of R has a minimal element with respect to the inclusion relation; an equivalent condition is that every infinite descending chain of left ideals of R should level off. We have similar conditions for R to be a left Noetherian ring; referring to Theorem 11.2 we see also that R is a left Noetherian ring if and only if every left ideal is finitely generated.

EXERCISES 2

1. Let W be any subset of a left R-module V; we define the annihilator of W to be the set

$$W^\flat = \{a \in R : ax = 0 \text{ for all } x \text{ in } W\}.$$

Prove that $W^\flat$ is a left ideal of R and that if W is a submodule of V then $W^\flat$ is a two-sided ideal of R.

A module V is said to be faithful if its annihilator consists of the zero element of R alone. Show that each left R-module V can be given the structure of a faithful left $(R/V^\flat)$-module.

2. For each left R-module V let $\gamma(V)$ be the least member of the set of cardinals of all generating systems for V. If

$$0 \to A' \to A \to A'' \to 0$$

is a short exact sequence of left R-modules, prove that

$$\gamma(A'') \leqslant \gamma(A) \leqslant \gamma(A') + \gamma(A'')$$

3. In the diagram

$$
\begin{array}{ccccc}
A & \xrightarrow{\;\varphi\;} & B & \xrightarrow{\;\psi\;} & C \\
\alpha \downarrow & & \beta \downarrow & & \gamma \downarrow \\
A' & \xrightarrow{\;\varphi'\;} & B' & \xrightarrow{\;\psi'\;} & C'
\end{array}
$$

of left R-modules and homomorphisms both squares are commutative and both rows are exact. Prove that $\operatorname{Im}\beta \cap \operatorname{Im}\varphi' = \beta(\operatorname{Ker}\gamma\psi)$ and $\operatorname{Ker}\beta + \operatorname{Ker}\psi = \beta^{-1}(\operatorname{Im}\varphi'\alpha)$.

4. In the diagram

$$
\begin{array}{ccccccc}
 & & B & \xrightarrow{\;\psi\;} & C & \xrightarrow{\;\theta\;} & D \\
 & & \beta \downarrow & & \gamma \downarrow & & \delta \downarrow \\
A' & \xrightarrow{\;\varphi'\;} & B' & \xrightarrow{\;\psi'\;} & C' & \xrightarrow{\;\theta'\;} & D' \\
\alpha' \downarrow & & \beta' \downarrow & & \gamma' \downarrow & & \\
A'' & \xrightarrow{\;\varphi''\;} & B'' & \xrightarrow{\;\psi''\;} & C'' & & \\
\alpha'' \downarrow & & \beta'' \downarrow & & & & \\
A''' & \xrightarrow{\;\varphi'''\;} & B''' & & & &
\end{array}
$$

of left R-modules and homomorphisms, all squares are commutative and all rows and columns are exact. If δ and φ'' are monomorphisms and α'' is an epimorphism, prove that φ''' is a monomorphism.

5. Let R and S be rings, A a left R-module, B a right S-module and C a left R-module which is also a right S-module such that $r(cs) = (rc)s$ for all elements r, c, s of R, C, S respectively. Show how to define the structure of a left R-module on $\operatorname{Hom}_S(B, C)$ and the structure of a right S-module on $\operatorname{Hom}_R(A, C)$. Prove that the abelian groups $\operatorname{Hom}_R(A, \operatorname{Hom}_S(B, C))$ and $\operatorname{Hom}_S(B, \operatorname{Hom}_R(A, C))$ are isomorphic.

6. Prove that the short exact sequence

$$0 \to A' \xrightarrow{\alpha} A \xrightarrow{\beta} A'' \to 0$$

of left R-modules and homomorphisms is split if and only if there exist homomorphisms θ from A'' to A and φ from A to A' such that $\alpha\varphi + \theta\beta = I_A$.

7. Let $(R_k)_{k \in K}$ be a family of subrings of a ring R such that the additive group R^+ of R is the internal direct sum of the family (R_k^+) of additive groups of the subrings. Prove that the following conditions are equivalent:

(a) for every pair of elements $x = \sum_{k \in K} x_k$ and $y = \sum_{k \in K} y_k$ of R we have $xy = \sum_{k \in K} x_k y_k$;

(b) each of the subrings R_k is a two-sided ideal of R;

(c) if i and j are distinct indices in K then for all elements x_i in R_i, x_j in R_j we have $x_i x_j = 0$.

8. Let A be an abelian group, R its ring of endomorphisms. If B is any abelian group show how to give $\operatorname{Hom}_Z(B, A)$ the structure of a left R-module. Prove that if B is a direct summand of A then $\operatorname{Hom}_Z(B, A)$ is a direct summand of R_l.

9. Let $(V_k)_{k \in K}$ be a family of left R-modules, V the external direct sum of this family. If W is any left R-module prove that $\operatorname{Hom}_R(V, W)$ is isomorphic to the direct product of the family $(\operatorname{Hom}_R(V_k, W))_{k \in K}$ of abelian groups.

10. In the diagram

$$
\begin{array}{ccccccccc}
& & P' & & & & P'' & & \\
& & \downarrow{\varphi'} & & & & \downarrow{\varphi''} & & \\
0 & \longrightarrow & A' & \xrightarrow{\alpha} & A & \xrightarrow{\beta} & A'' & \longrightarrow & 0
\end{array}
$$

of left R-modules and homomorphisms P' and P'' are projective and the row is exact. Show that there exist a projective module P and homomorphisms ι', π'' and φ such that in the diagram

$$
\begin{array}{ccccccccc}
0 & \longrightarrow & P' & \xrightarrow{\iota'} & P & \xrightarrow{\pi''} & P'' & \longrightarrow & 0 \\
 & & \varphi' \downarrow & & \varphi \downarrow & & \downarrow \varphi'' & & \\
0 & \longrightarrow & A' & \xrightarrow{\alpha} & A & \xrightarrow{\beta} & A'' & \longrightarrow & 0
\end{array}
$$

both rows are exact and both squares are commutative. Show further that if φ' and φ'' are epimorphisms so is φ.

11. Prove that the direct product of a family of injective modules is injective and the external direct sum of a family of projective modules is projective.

12. Prove that a left R-module Q is injective if and only if for every projective left R-module P, every submodule P' of P and every homomorphism φ' from P' to Q there exists an extension φ of φ' from P to Q.

13. Let R be a ring such that every submodule of every projective left R-module is projective. Prove that every quotient module of every injective R-module is injective.

14. Let A be an abelian group, R a ring, R^+ its additive group. Show how to give $\text{Hom}_Z(R^+, A)$ the structure of a left R-module. Prove that if A is an injective $\mathbf{Z}$-module then $\text{Hom}_Z(R^+, A)$ is an injective R-module.

15. Let A be an abelian group, n a positive integer, v_A the endomorphism of A defined by setting $v_A(a) = na$ for all elements a of A; write $nA = \text{Im } v_A$. By considering the exact sequence

$$
0 \longrightarrow \mathbf{Z} \xrightarrow{v_Z} \mathbf{Z} \longrightarrow \mathbf{Z}/n\mathbf{Z} \longrightarrow 0
$$

prove that for all abelian groups A we have $A \otimes_Z (\mathbf{Z}/n\mathbf{Z})$ isomorphic to A/nA.

16. Let R be a ring and let

$$
\begin{array}{ccccccccc}
0 & \longrightarrow & A' & \xrightarrow{\alpha_1} & A & \xrightarrow{\alpha_2} & A'' & \longrightarrow & 0, \\
0 & \longrightarrow & B' & \xrightarrow{\beta_1} & B & \xrightarrow{\beta_2} & B'' & \longrightarrow & 0
\end{array}
$$

be short exact sequences of right and left R-modules respectively. Prove that $\alpha_2 \otimes \beta_2$ is an epimorphism from $A \otimes_R B$ onto $A'' \otimes_R B''$ and that

$$\text{Ker}\,(\alpha_2 \otimes \beta_2) = \text{Im}\,\alpha_1 \otimes I_{B'} + \text{Im}\,I_{A'} \otimes \beta_1.$$

17. Let V and W be modules over a commutative ring R, V^* and W^* their duals. Prove that there exists a unique abelian group homomorphism λ from $V^* \otimes_R W^*$ to $(V \otimes_R W)^*$ such that

$$[\lambda(\alpha \otimes \beta)]\,(x \otimes y) = \alpha(x)\beta(y)$$

for all elements α, β, x, y of V^*, W^*, V, W respectively. If R is a field and V and W are finite-dimensional vector spaces over R show that λ is an isomorphism.

18. Let A_1, A_2 be ideals of a commutative ring R. Prove that $(R/A_1) \otimes_R (R/A_2)$ is isomorphic to $R/(A_1 + A_2)$.

CHAPTER 3
COMMUTATIVE RINGS

§ 12. Unique Factorisation Domains

One of the important properties of the ring of integers $\mathbf{Z}$, which most of us learned in a more or less informal way at an early age, is that every integer can be expressed in an essentially unique way as a product of prime numbers. In this section we shall try to express this property in terms which might be applicable to more general rings than $\mathbf{Z}$, and shall then produce some conditions under which rings will enjoy the property.

We restrict ourselves to integral domains with identity, i.e. to commutative rings with identity having no divisors of zero.

So let R be an integral domain, with identity element e. If a and b are elements of R then we say that a is a *multiple* of b, that b is *divisible* by a and that b is a *factor* or *divisor* of a if there exists an element c of R such that $a = bc$. (Since R is commutative it is unnecessary to distinguish between right and left multiples or right and left factors.) In this situation we write $b|a$ and read this as 'b *divides* a'.

We recall that an invertible element of R is called a *unit*, and we easily verify that an element is a unit if and only if it divides the identity element e. If u is a unit of R with inverse u' and a is any element of R, we have $a = ea = (uu')a = u(u'a)$; thus the units of R are factors of every element of R.

If a and b are elements of R such that $a|b$ and $b|a$ we say that a and b are *associates* and we write $a \sim b$. It is easy to see that a and b are associates if and only if there is a unit u of R such that $a = bu$. For, if there is such a unit u, with inverse u' say, we have $au' = (bu)u' = b(uu') = be = b$; hence $a|b$. Since $a = bu$ we have $b|a$; thus $a \sim b$. Conversely, if $a \sim b$ there are elements u, v of R

94

such that $a = bu$ and $b = av$. Then we have $a = bu = (av)u = a(vu)$; hence $a(e - vu) = 0$. Since R is an integral domain we have either $a = 0$, whence $b = 0 = 0e = ae$, or else $e - vu = 0$, whence $vu = e$ and u is a unit. It is clear that the relation $x \sim y$ is an equivalence relation on R.

If $b \mid a$ and b is neither a unit nor an associate of a, we say that b is a *proper factor* of a. An element of R is said to be *irreducible* if it is not a unit and has no proper factors. Clearly if p is irreducible and $p \sim p'$ then p' is also irreducible.

Since R is a commutative ring with identity it follows from our discussions in §3 that the principal ideal (a) generated by the element a of R consists of all the multiples of a. We can thus reformulate the preceding discussion in terms of principal ideals as follows.

THEOREM 12.1. *Let R be an integral domain with identity. An element u of R is a unit if and only if $(u) = R$. If a and b are elements of R then b is a factor of a if and only if $(a) \subseteq (b)$, and b is a proper factor of a if and only if $(a) \subset (b) \neq R$. Further, a and b are associates if and only if $(a) = (b)$.*

Proof. (1) We have already shown that if u is a unit then every element of R is a multiple of u; so $(u) = R$. Conversely, if $(u) = R$, then every element of R is a multiple of u and so, in particular, there is an element v of R such that $e = uv$, i.e. u is a unit.

(2) Suppose $b \mid a$; then there is an element c of R such that $a = bc$. If x is any element of (a) there is an element x' of R such that $x = ax'$. Then $x = (bc)x' = b(cx') \in (b)$; so $(a) \subseteq (b)$. Conversely, if $(a) \subseteq (b)$, then $a \in (b)$ and so a is a multiple of b.

(3) It follows from (2) that $a \sim b$ (i.e. $a \mid b$ and $b \mid a$) if and only if $(a) \subseteq (b)$ and $(b) \subseteq (a)$, i.e. if and only if $(a) = (b)$.

(4) If b is a proper factor of a, then according to (2) we have $(a) \subseteq (b)$. But, since b is not an associate of a, (3) shows that we cannot have $(a) = (b)$; and since b is not a unit it follows from (1) that $(b) \neq R$. Hence we have $(a) \subset (b) \neq R$. The converse is obtained by reversing the steps of the argument.

Let a be a non-zero element of the integral domain R which is not a unit of R. Then a *factorisation* of a is a finite family $(p_i)_{i \in I}$ of irreducible elements of R such that $a = \prod_{i \in I} p_i$. We say that R is a *factorisation domain* if every non-zero non-unit element of R has at least one factorisation. Our next theorem gives a sufficient condition for R to be a factorisation domain.

THEOREM 12.2. *Let R be an integral domain. If there exist no infinite sequences $(a_i)_{i \in}$ of elements of R such that a_{i+1} is a proper factor of a_i (for all natural numbers i), then R is a factorisation domain.*

Proof. Let a be any non-zero non-unit element of R.

Set $a_0 = a$. Now suppose that for some natural number k we have defined elements $a_0, a_1, \ldots, a_k$ of R such that for all natural numbers $i < k$ the element a_{i+1} is a proper factor of a_1. If a_k is not irreducible it has a proper factor; we choose one of its proper factors and call it a_{k+1}. It is clear that we must eventually reach an irreducible element a_m, since otherwise we would have an infinite sequence $(a_i)_{i \in \mathbf{N}}$ such that each a_{i+1} is a proper factor of a_i; and this contradicts our hypothesis about R.

We now set $a'_0 = a$, $p_0 = a_m$. Since p_0 is an irreducible factor of a'_0, there is an element a'_1 such that $a'_0 = p_0 a'_1$. Now suppose that for some natural number k we have defined irreducible elements $p_0, \ldots, p_k$ of R and elements $a'_0, a'_1, \ldots, a'_{k+1}$ of R such that $a'_i = p_i a'_{i+1}$ for $i = 0, \ldots, k$. If a'_{k+1} is not a unit the argument above shows that a'_{k+1} has an irreducible factor, p_{k+1} say, and there is an element a'_{k+2} such that $a'_{k+1} = p_{k+1} a'_{k+2}$. It is clear that when we repeat this procedure we must eventually come to a unit a'_{n+1}, since otherwise we would have an infinite sequence $(a'_i)_{i \in \mathbf{N}}$ such that a'_{i+1} is a proper factor of a'_i, again contradicting our hypothesis about R.

Without loss of generality, we may assume that $a'_{n+1} = e$. (If p_n is irreducible and a'_{n+1} is a unit, then $p_n a'_{n+1}$ is also irreducible.) Then we have $a = \prod_{i=0}^{n} p_i$, i.e. $(p_i)_{i \in [0, n]}$ is a factorisation of a.

This completes the proof.

Referring to Theorem 12.1 we can give an alternative criterion for R to be a factorisation domain.

COROLLARY. *Let R be an integral domain. If there are no infinite strictly ascending chains of principal ideals of R then R is a factorisation domain. In particular, every Noetherian integral domain is a factorisation domain.*

The existence of a factorisation of an element of a domain is not by itself particularly interesting or important; it becomes interesting only when the factorisation is 'essentially unique'. To explain the meaning of this phrase, let us suppose that we have a factorisation $(p_i)_{i \in I}$ of an element a of the domain R. If $(u_i)_{i \in I}$ is a family of units of R with the same index set I such that $\prod_{i \in I} u_i = e$, the elements $u_i p_i$ are all irreducible and $\prod_{i \in} u_i p_i = a$; so $(u_i p_i)_{i \in I}$ is another factorisation of a. Again, if π is a bijection from I onto a set J and for each index j in J we set $q_j = p_{\pi^{-1}(j)}$, it follows from the commutativity of R that $\prod_{j \in J} q_j = \prod_{i \in I} p_i = a$; so $(q_j)_{j \in J}$ is another factorisation of a. If all factorisations of a are obtained from $(p_i)_{i \in I}$ by a succession of transformations of these two 'inessential' types, we say that the factorisation of a is essentially unique; formally, a has an *essentially unique factorisation* if and only if whenever $(p_i)_{i \in I}$ and $(q_j)_{j \in J}$ are factorisations of a, then there is a bijection π from I onto J such that for each index i in I we have $p_i \sim q_{\pi(i)}$. An integral domain is called a *unique factorisation domain* if every non-zero non-unit of R has an essentially unique factorisation. Our next concern is to discover conditions under which a factorisation domain is actually a unique factorisation domain.

For our first theorem in this direction we need the notion of a prime element of a ring; we say that the element p is *prime* if it is not a unit and whenever a product ab of elements of the ring is divisible by p then either $p|a$ or $p|b$. It is easy to see that every prime element is irreducible. For if p is a prime element, then (by definition) it is not a unit; we shall show that it has no proper factors. So suppose a is a factor of p, say $p = ab$; since $p|p$ it follows that either $p|a$ or $p|b$. If $p|a$ we have $a \sim p$; so a is not a proper factor of p; if $p|b$ we have $b \sim p$ and hence a is a unit—so again a is not a proper factor of p. Thus p is irreducible. We now show that the converse of this property is a criterion for uniqueness of factorisation.

THEOREM 12.3. *Let R be a factorisation domain. Then R is a unique factorisation domain if and only if every irreducible element of R is prime.*

Proof. (1) Suppose every irreducible element of R is prime.

We make the inductive hypothesis that for every element of R which has a factorisation $(p_i)_{i \in I}$ with Card $I < n$ this factorisation is essentially unique.

Now let a be an element with a factorisation $(p_i)_{i \in I}$ such that

Card $I = n$. We shall show that this factorisation is essentially unique. We may assume without loss of generality that $I = [1, n]$; then $a = \prod\limits_{i \in 1}^{n} p_i$. Suppose we have another factorisation $(q_j)_{j \in [1, m]}$ for a, so that $a = \prod\limits_{j=1}^{m} q_j$.

Since $a = (\prod\limits_{i=1}^{n-1} p_i)p_n$ it follows that p_n is a factor of the product $\prod\limits_{j=1}^{m} q_j$. The element p_n is irreducible, and hence prime; so p_n must divide one of the factors in the family (q_j)—this follows by a simple inductive argument based on the definition of a prime element; say $p_n | q_k$. Since q_k is irreducible it follows that $p_n \sim q_k$, say $q_k = up_n$, where u is a unit.

Define a bijection π from $[1, m]$ onto itself by setting $\pi(m) = k$, $\pi(k) = m$ and $\pi(j) = j$ for other indices j in $[1, m]$. Then $a = \prod\limits_{j=1}^{m} q_{\pi(j)} = (\prod\limits_{j=1}^{m-1} q_{\pi(j)}) q_k = (\prod\limits_{j=1}^{m-1} q_{\pi(j)}) up_n = (\prod\limits_{j=1}^{m-1} q'_j) p_n$, where $q'_1 = uq_{\pi(1)}$ and $q'_j = q_{\pi(j)}$ for $j = 2, \ldots, m - 1$. Hence

$$(\prod\limits_{i=1}^{n-1} p_i) p_n = (\prod\limits_{j=1}^{m-1} q'_j) p_n$$

and so, since R is an integral domain, we have

$$\prod\limits_{i=1}^{n-1} p_i = \prod\limits_{j=1}^{m-1} q'_j.$$

We now apply the inductive hypothesis and deduce first that $n - 1 = m - 1$ (whence $n = m$) and then that there is a bijection κ from $[1, n - 1]$ onto itself such that for $i = 1, \ldots, n - 1$ we have $p_i \sim q'_{\kappa(i)}$ and so, of course, $p_i \sim q_{\pi\kappa(i)}$.

Consider now the mapping λ from $[1, n]$ to itself defined by setting $\lambda(i) = \pi\kappa(i)$ for $i = 1, \ldots, n - 1$ and $\lambda(n) = k$. It is easily verified that λ is a bijection and clearly $p_i \sim q_{\lambda(i)}$ for $i = 1, \ldots, n$.

This completes the induction; hence every element of R has an essentially unique factorisation, as required.

(2) Conversely, suppose R is a unique factorisation domain.

Let p be an irreducible element of R and suppose p is a factor of the product ab of two elements a, b of R. If $a = 0$ then $p|a$ and similarly if $b = 0$ then $p|b$; again, if a is a unit then $ab \sim b$, so that

$p|b$ and similarly if b is a unit then $p|a$. So we may suppose that a and b are non-zero non-units of R. Since $p|ab$ there is an element c of R such that $ab = pc$. Clearly c is non-zero, and c is not a unit since p is irreducible.

Let $(p_i)_{i \in I}$, $(p'_j)_{j \in J}$, $(q_k)_{k \in K}$ be factorisations for a, b, c respectively. Then $p \prod_{k \in K} q_k = (\prod_{i \in I} p_i)(\prod_{i \in J} p'_j)$. It follows from the uniqueness of factorisation that the irreducible element p, which occurs in the factorisation on the left, is associated with one of the irreducible elements in the factorisation on the right. If p is associated with one of the elements p_i then $p|a$; similarly, if p is associated with one of the elements p'_j then $p|b$.

Let a and b be elements of a ring R. An element d of R is called a *highest common factor* or *greatest common divisor* of a and b, if (1) $d|a$ and $d|b$ and (2) for every element c of R such that $c|a$ and $c|b$ we have $c|d$. It is clear at once that if d_1 and d_2 are highest common factors of a and b then d_1 and d_2 are associates, and conversely, if d is a highest common factor of a and b then so is every associate of d. If the identity element e is a highest common factor of a and b we say that a and b are *relatively prime*. We remark that it is not true in general that every pair of elements in a ring has a highest common factor; in fact we shall show that this property is another criterion for uniqueness of factorisation.

THEOREM 12.4. *Let R be a factorisation domain. Then R is a unique factorisation domain if and only if every pair of non-zero elements of R has a highest common factor.*

Proof. (1) Suppose every pair of non-zero elements of R has a highest common factor.

Let p be an irreducible element of R; we shall show that p is prime. So let a and b be elements of R such that $p|ab$. We want to show that either a or b is divisible by p, or, in other words, that if a is not divisible by p then $p|b$. Since $p|0$ we may assume that b is non-zero.

Suppose then that p does not divide a. Since p is irreducible its only factors are the units of R and the associates of p; since p is not even a factor of a it cannot be a highest common factor of a and p. Hence the units of R are the only common factors of a and p; in particular, the identity element e is a highest common factor.

Now we claim that b is a highest common factor of ab and pb. To see this, let b' be any highest common factor of ab and pb. Since

$b \mid ab$ and $b \mid pb$ we certainly have $b \mid b'$, and so there is an element c of R such that $b' = cb$. Since $b' \mid ab$ and $b' \mid pb$ there are elements x and y of R such that $ab = xb' = xcb$ and $pb = yb' = ycb$. Because R is an integral domain we deduce that $a = xc$, $p = yc$; so c is a common factor of a and p and hence c divides the highest common factor e of these elements. Hence c is a unit, $b \sim b'$ and so b is a highest common factor of ab and pb.

Since $p \mid ab$ by hypothesis and obviously $p \mid pb$, it follows that $p \mid b$, as required.

Hence every irreducible element of R is prime, and it follows from Theorem 12.3 that R is a unique factorisation domain.

(2) Conversely, let R be a unique factorisation domain.

Let P be the set of all equivalence classes under the equivalence relation $x \sim y$ which consist of irreducible elements of R; from each class C in P choose a representative p_C, so that $p \in C$ if and only if $p \sim p_C$.

Let a and b be non-zero elements of R. If either a or b is a unit then the identity element e is a highest common factor of a and b; so we may assume that a and b are both non-units.

If $(p_i)_{i \in I}$ is a factorisation for a then for each class C in P we write $\mu(C) = \text{Card } \{i \in I \mid p_i \in C\} = $ the number of irreducible factors in the family $(p_i)_{i \in I}$ which belong to the class C. Since I is finite we have $\mu(C) \neq 0$ for at most finitely many classes C; thus the product $\prod_{C \in P} p_C^{\mu(C)}$ is essentially finite (the factors of the form p_C^0 may all be dropped) and clearly we have $a = \prod_{i \in I} p_i \sim \prod_{C \in P} p_C^{\mu(C)}$. In the same way, there are natural numbers $v(C)$ such that $b \sim \prod_{C \in P} p_C^{v(C)}$.

For each equivalence class C set $\lambda(C) = \inf \left(\mu(C), v(C)\right)$ and write $d = \prod_{C \in P} p_C^{\lambda(C)}$. Then it is easy to verify that d is a highest common factor of a and b.

Let $(a_i)_{i \in I}$ be a finite family of elements in a ring R; then an element d of R is called a *highest common factor* or *greatest common divisor* of the family (a_i) if (1) for each index i in I, $d \mid a_i$ and (2) if c is any element of R such that $c \mid a_i$ for each index i then $c \mid d$. It is clear that if d_1 and d_2 are two highest common factors of (a_i) then d_1 and d_2 are associates. We now have the following consequence of Theorem 12.4.

COROLLARY. *Let R be a unique factorisation domain. Then every finite family of elements of R has a highest common factor.*

Proof. Let $(a_i)_{i \in I}$ be a finite family of elements of R. Without loss of generality we may take I to be the interval $[1, n]$ of $\mathbf{N}$. We proceed by induction on n. If $n = 1$ the result is clear. So we make the inductive hypothesis that for some natural number k we have proved that all finite families $(a_i)_{i \in I}$ with Card $I < k$ have highest common factors.

Now consider a family $(a_i)_{i \in [1, k]}$ where $k > 1$. By hypothesis, the family $(a_i)_{i \in [1, k-1]}$ has a highest common factor, a say. According to Theorem 12.4, there is a highest common factor d of a and a_k. We claim that d is a highest common factor of $(a_i)_{i \in [1, k]}$.

Since $d \mid a$ and $a \mid a_i$ $(i = 1, \ldots, k - 1)$ it follows that $d \mid a_i$ $(i = 1, \ldots, k - 1)$; but of course $d \mid a_k$ also. So $d \mid a_i$ $(i = 1, \ldots, k)$. Next suppose $c \mid a_i$ $(i = 1, \ldots, k)$. Then c divides each highest common factor of $a_1, \ldots, a_{k-1}$, so $c \mid a$; and $c \mid a_k$. Hence c divides each common factor of a and a_k, so $c \mid d$.

Thus d is a highest common factor of $(a_i)_{i \in [1, k]}$.

A commutative ring in which every ideal is principal is called a *principal ideal ring*; an integral domain in which every ideal is principal is called a *principal ideal domain*.

THEOREM 12.5. *Every principal ideal domain is a unique factorisation domain.*

Proof. Let R be a principal ideal domain.

Since every ideal of R is finitely generated (actually by a single element), R is Noetherian and hence, by the Corollary to Theorem 12.2, R is a factorisation domain.

To show that it is a unique factorisation domain, we apply the criterion of Theorem 12.4. So let a and b be non-zero elements of R. Since R is a principal ideal ring there is an element d of R such that the ideal sum $(a) + (b) = (d)$. We claim that d is a highest common factor of a and b. Since $a = a + 0 \in (a) + (b) = (d)$ it follows that $d \mid a$; similarly $d \mid b$. On the other hand, since $d \in (d) = (a) + (b)$, there are elements x and y of R such that $d = ax + by$; hence if $c \mid a$ and $c \mid b$ we must have $c \mid d$. So d is indeed a highest common factor of a and b.

The desired result now follows from Theorem 12.4.

We introduce now an interesting subclass of the class of principal ideal domains. Let R be an integral domain. A *Euclidean norm* on R is a mapping v from the set of non-zero elements of R to the set of natural numbers $\mathbf{N}$ such that the following conditions are satisfied.

E1. If a and b are non-zero elements of R such that $a|b$ we have $v(b) \geqslant v(a)$.

E2. For every element a of R and every non-zero element d of R there exist elements q and r such that $a = qd + r$ and either $r = 0$ or else $v(r) < v(d)$.

An integral domain which is equipped with a Euclidean norm v is called a *Euclidean domain* with norm v.

Example 1. Let v be the mapping from the set of non-zero integers to $\mathbf{N}$ defined by setting $v(a) = |a| = \sup (a, -a)$ for all non-zero integers a. We claim that v is a Euclidean norm on $\mathbf{Z}$.

If a and b are non-zero integers such that $a|b$, there is a non-zero integer c such that $b = ac$. Then $|c| \geqslant 1$ and so $v(b) = |ac| = |a||c| \geqslant |a| = v(a)$. So v satisfies E1.

Next let d be a non-zero integer. If $a = 0$, we can write $a = 0d + 0$. Next let d be positive, and suppose first that a is positive. Since $(a + 1)d \geqslant a + 1 > a$ the set A of natural numbers k such that $kd > a$ is non-empty, and hence has a least element m say. Clearly m is non-zero; so there is a natural number q such that $m = q + 1$. Then we have $qd \leqslant a < (q + 1)d$; so $0 \leqslant a - qd < d$. Hence if $r = a - qd$ we have $a = qd + r$ and either $r = 0$ or $v(r) = |r| < d = v(d)$. If a is negative then $-a$ is positive and the preceding discussion shows that there are integers q_1, r_1 such that $-a = q_1 d + r_1$ and r_1 is either zero or else $v(r_1) = r_1 < v(d)$. Then $a = (-q_1)d + (-r_1)$ and $-r_1$ is either zero or else $v(-r_1) = |-r_1| = |r_1| < v(d)$. Finally let d be negative, a any non-zero integer. Then $-d$ is positive, and hence there are integers q_1, r_1 such that $a = q_1(-d) + r_1 = (-q_1)d + r_1$ and r_1 is either zero or else $v(r_1) < v(-d) = v(d)$. Hence v satisfies E2.

Example 2. Let $\mathbf{G}$ be the set of complex numbers of the form $x + yi$ where x and y are integers; it is easy to verify that $\mathbf{G}$ is a subring of the field of complex numbers $\mathbf{C}$. We call $\mathbf{G}$ the ring of *Gaussian integers*. Let v be the mapping from the set of non-zero elements of $\mathbf{G}$ to $\mathbf{N}$ defined by setting $v(x + yi) = x^2 + y^2 = |x + yi|^2$ for every non-zero element $x + yi$ of $\mathbf{G}$. We shall show that v is a Euclidean norm on $\mathbf{G}$.

If a and b are non-zero elements of $\mathbf{G}$ such that $a|b$, there is a non-zero Gaussian integer c such that $b = ac$. Then $|c| \geqslant 1$ and so we have $v(b) = |b|^2 = |a|^2 |c|^2 \geqslant |a|^2 = v(a)$. So v satisfies $E1$.

Now, let a be any element of $\mathbf{G}$, d any non-zero element. Since $\mathbf{G}$ is included in $\mathbf{C}$ and $\mathbf{C}$ is a field, we can form the element $q' = ad^{-1} = x' + y'i$ of $\mathbf{C}$. (In general x' and y' will be rational numbers, not integers.) There are integers x and y such that $|x - x'| \leqslant \frac{1}{2}, |y - y'| \leqslant \frac{1}{2}$. Let $q = x + yi$; then we have $v(q' - q) = |q' - q|^2 = |x - x'|^2 + |y - y'|^2 \leqslant \frac{1}{2}$. Now set $r = a - qd = (q' - q)d$. If $q' = q$ we have $r = 0$; otherwise $v(r) = |(q' - q)d|^2 = |q' - q|^2 |d|^2 \leqslant \frac{1}{2}|d|^2 < v(d)$. Thus v satisfies $E2$.

Example 3. Let F be a field, $R = P(F)$ the ring of polynomials with coefficients in F. Let v be the mapping from the set of non-zero polynomials of R to $\mathbf{N}$ defined by setting $v(f) = \partial f (= $ the degree of $f)$ for every non-zero polynomial f in R. We shall show that v is a Euclidean norm on R.

If a and b are non-zero polynomials in R such that $a|b$ there is a polynomial c in R such that $b = ac$. Then $v(b) = \partial b = \partial(ac) = \partial a + \partial c \geqslant \partial a$. (Note that we have the equality $\partial(ac) = \partial a + \partial c$ since F is a field.) Thus v satisfies $E1$.

Let a be any polynomial in R, d any non-zero polynomial in R. If $a = z$, the zero polynomial, we have $a = zd + z$. So we may assume that a is not the zero polynomial. If $\partial a < \partial d$ we have $a = zd + a$. So we now suppose that $\partial a \geqslant \partial d$ and proceed by induction on the degree of a. Thus we let $\partial a = n \geqslant \partial d = m$ and make the inductive hypothesis that for all polynomials a_1 such that $\partial a_1 < n$ there exist polynomials q_1 and r_1 such that $a_1 = q_1 d + r_1$ and either $r_1 = z$ or else $\partial r_1 < \partial d$. Suppose $a = a_n X^n + \ldots + a_0$, $d = d_m X^m + \ldots + d_0$. Set $a_1 = a - a_n d_m^{-1} X^{n-m} d$; then $\partial a_1 < \partial a$ and hence there are polynomials q_1, r_1 of the type just described. Hence, by setting $q = q_1 + a_n d_m^{-1} X^{n-m} d$ and $r = r_1$ we have $a = qd + r$ and either $r = z$ or $\partial r < \partial d$. Thus v satisfies $E2$.

We now prove the main theorem on Euclidean domains.

THEOREM 12.6. *Every Euclidean domain is a principal ideal domain.*

Proof. Let R be a Euclidean domain, I any ideal of R.

Since the zero ideal is principal, we may suppose that I is non-zero. If I^* is the set of non-zero elements of I, $v(I^*)$ is a non-empty subset of N and hence has a least element. That is to say, there is a non-zero element d of I such that for all elements x of I we have $v(d) \leqslant v(x)$. We shall show that I is the principal ideal generated by d.

Clearly $(d) \subseteq I$. So let a be any element of I. Since v is a Euclidean norm, there exist elements q and r of R such that $a = qd + r$ and either $r = 0$ or else $v(r) < v(d)$. Since $a \in I$ and $qd \in I$ it follows that $r = a - qd$ belongs to I. Hence r must be zero, for otherwise we would have a contradiction to the minimal property of $v(d)$. Thus $a = qd \in (d)$ and so $I = (d)$, as asserted.

COROLLARY. *Every Euclidean domain is a unique factorisation domain.*

§ 13. Maximal Ideals and Prime Ideals

Let R be a commutative ring; in R the concepts of left, right and two-sided ideals coincide, so we may talk simply of ' ideals ' in R. According to the definition which we gave in §3, an ideal M of R is maximal if and only if (1) $M \neq R$ and (2) there is no ideal I of R such that $M \subset I \subset R$. We now give a criterion for an ideal in a commutative ring with identity to be maximal.

THEOREM 13.1. *Let R be a commutative ring with an identity element e. Then an ideal M of R is maximal if and only if the residue class ring R/M is a field.*

Proof. (1) Suppose M is a maximal ideal of R. We have already remarked in §3 that if R is a commutative ring with identity then so is every residue class ring. Hence all we have to prove is that every non-zero element of R/M has a multiplicative inverse.

So let C be any non-zero element of R/M; choose any element a of R such that $C = \eta(a)$, where η is the canonical epimorphism from R onto R/M. Since C is not the zero element of R/M, a does not belong to M. Consider now the ideal I of R generated by $M \cup \{a\}$; it is easy to check that I consists of all elements of the form $m + ra$ such that $m \in M$ and $r \in R$. Since $M \subseteq I$, $a \in I$ and $a \notin M$ it follows that $M \subset I$ and hence, since M is maximal, we have $I = R$. We deduce in particular that the identity element e of R belongs to I; in other words, there exist elements m_0 of M and a' of R such that $e = m_0 + a'a$. Applying the epimorphism η, we have that $\eta(e) = \eta(m_0 + a'a) = \eta(a'a) = \eta(a')\eta(a) = \eta(a')C$; so $\eta(a')$ is a multiplicative inverse for C.

Hence R/M is a field.

(2) Conversely, suppose R/M is a field.

Let I be any ideal of R such that $I \supset M$; choose any element a of I which does not belong to M. Then $\eta(a)$ is not the zero element of R/M and hence $\eta(a)$ has a multiplicative inverse in R/M, i.e. there exists an element C' of R/M such that $\eta(a)C' = \eta(e)$. If a' is any element of this residue class C' we have $\eta(e) = \eta(a)\eta(a') = \eta(aa')$ and so there is an element m of M such that $e = m + aa'$. Since $m \in M \subset I$ and $a \in I$ it follows that $e \in I$ and hence $I = R$.

Thus M is a maximal ideal of R.

Now, let R be any commutative ring, not necessarily with an identity. An ideal P of R is said to be a *prime ideal* if the residue class ring R/P is an integral domain. Since every field is an integral domain it follows from Theorem 13.1 that every maximal ideal in a commutative ring with identity is prime. In the next two theorems we give criteria for an ideal in an arbitrary commutative ring to be prime.

THEOREM 13.2. *Let R be a commutative ring. Then an ideal P of R is prime if and only if whenever the product ab of two elements of R belongs to P then at least one of the elements a, b belongs to P.*

Proof. (1) Suppose P is a prime ideal of R.

Let a and b be elements of R such that $ab \in P$. Then $\eta(ab) = \eta(a)\eta(b)$ is the zero element of R/P (where η is the canonical epimorphism onto R/P). Since R/P is an integral domain, it has no divisors of zero; so either $\eta(a)$ or $\eta(b)$ is the zero element of R/P. Thus either $a \in P$ or $b \in P$, as required.

(2) Conversely, suppose that the condition of the theorem is satisfied.

Let A and B be elements of R/P (residue classes modulo P) such that $AB = \eta(0)$. Choose elements a, b of R from A and B respectively; then we have $\eta(ab) = \eta(a)\eta(b) = AB = \eta(0)$ and hence $ab \in P$. It follows from our hypothesis that either $a \in P$ or $b \in P$; hence either $A = \eta(a) = \eta(0)$ or $B = \eta(b) = \eta(0)$.

Thus R/P has no divisors of zero. Since R is commutative, so is R/P, and hence R/P is an integral domain, i.e. P is a prime ideal as asserted.

COROLLARY. *Let α be an epimorphism from a commutative ring R onto* *another commutative ring R'. If P is any prime ideal of R which includes the kernel of α then $\alpha(P)$ is a prime ideal of R'.*

Proof. Let a' and b' be elements of R' such that $a'b' = \alpha(p)$ where $p \in P$. Since α is an epimorphism there are elements a, b of R such that $a' = \alpha(a)$ and $b' = \alpha(b)$. Then we have $\alpha(ab) = \alpha(a)\alpha(b) = a'b' = \alpha(p)$. It follows that $ab - p \in \operatorname{Ker} \alpha \subseteq P$ and hence $ab \in P$. Since P is prime we conclude that either $a \in P$ or $b \in P$, whence either $a' \in \alpha(P)$ or $b' \in \alpha(P)$. Thus $\alpha(P)$ is a prime ideal.

THEOREM 13.3. *Let R be a commutative ring. Then an ideal P of R is prime if and only if whenever the product IJ of two ideals of R is included in P then at least one of the ideals I, J is included in P.*

Proof. (1) Suppose P is a prime ideal of R.

Let I and J be ideals of R such that $IJ \subseteq P$. Suppose that neither I nor J is included in P. Then there exist elements a, b such that $a \in I$, $a \notin P$, $b \in J$, $b \notin P$. Since $ab \in IJ \subseteq P$, this contradicts our hypothesis that P is prime (we use here the criterion of Theorem 13.2).

Hence either $I \subseteq P$ or $J \subseteq P$.

(2) Conversely, suppose the condition of the theorem is satisfied, but that P is not prime. Then, according to Theorem 13.2, there are elements a, b of R such that $ab \in P$ but $a \notin P$ and $b \notin P$. Let A, B be the ideals of R generated by $P \cup \{a\}$, $P \cup \{b\}$ respectively. We deduce at once that $AB \subseteq P$ but $A \supset P$ and $B \supset P$, and this contradicts our hypothesis. Hence P is prime.

Example 1. If R is a commutative ring, R is itself a prime ideal of R. The zero ideal (0) is a prime ideal if and only if R is an integral domain.

Example 2. Let R be a unique factorisation domain. If u is a unit then the principal ideal (u) is the whole ring R and so is a prime ideal; if p is an irreducible element then it follows from Theorem 12.3 that p is a prime element and we deduce easily that (p) is a prime ideal. Conversely, suppose a is an element of R such that the principal ideal (a) is prime. If $(a) = R$ then a is a unit. If $(a) \neq R$, we claim that a is irreducible; it is certainly not a unit so we have only to show it has no proper factor. Suppose then that $a = bc$; we have $bc \in (a)$, and hence, since (a) is prime, either $b \in (a)$ or $c \in (a)$, i.e. either $a|b$ or $a|c$—so one of the factors b, c is an associate of a and the other is a unit. Thus in a unique factorisation domain the principal ideal (a) is a prime ideal if and only if a is either a unit or an irreducible element.

Example 3. Let R be a principal ideal domain. We shall show that every proper prime ideal is maximal. So let P be a proper prime ideal; since R is a principal ideal domain there is an element p such that $P = (p)$. Now R is a unique factorisation domain (Theorem 12.5) and hence, by Example 2 above, p is an irreducible element (it is not a unit, since P is a proper ideal). Let I be an ideal of R properly including P; let a be any element of I such that $a \notin P$. As in the proof of Theorem 12.5 it follows that a and p have a highest common factor d which can be expressed in the form $d = xa + yp$ where x and y are elements of R. But since p is irreducible and a is not divisible by p, the identity element e is a highest common factor of a and p. Hence $e \sim d$, i.e. $e = ud$ for a suitable unit u; so we have $e = uxa + uyp \in I$. Thus $I = R$ and P is maximal, as asserted.

Let I be an ideal in a commutative ring R; then the intersection of the family of prime ideals of R which include I is called the *radical* of I and is denoted by Rad I. It follows from the Corollary to Theorem 3.3 that Rad I is an ideal of R which includes I. We now give another characterization of Rad I.

THEOREM 13.4. *Let I be an ideal of a commutative ring R. Then the radical of I consists of all elements of R some power of which belongs to I.*

Proof. (1) Let a be an element of R such that some power of a, say a^k, belongs to I. If P is any prime ideal including I we have $a^k \in P$ and hence (using an inductive argument based on Theorem 13.2) $a \in P$. So $a \in$ Rad I.

(2) Conversely, let b be an element of R such that no power of b belongs to I. We shall show that there exists a prime ideal which does not contain b.

To this end let E be the set of ideals of R which include I but contain no power of b. Since I belongs to E, E is non-empty; a familiar argument shows that E is inductively ordered by the inclusion relation. Hence, according to Zorn's Lemma, E has a maximal element, P_0 say. We contend that P_0 is a prime ideal. So suppose x and y are elements of R such that $x \notin P_0$ and $y \notin P_0$; we shall prove that $xy \notin P_0$.

Since $x \notin P_0$ the ideal $P_0 + (x)$ properly includes P_0 and hence does not belong to the set E; hence there is a power of b, say b^k, which belongs to $P_0 + (x)$. So $b^k = p_1 + n_1 x + c_1 x$, where $p_1 \in P_0$, $n_1 \in \mathbf{Z}$ and $c_1 \in R$. Similarly since $y \notin P_0$ there is a power b^l of b which can be expressed in the form $b^l = p_2 + n_2 y + c_2 y$ with $p_2 \in P_0$, $n_2 \in \mathbf{Z}$, $c_2 \in R$. Hence we have

$$b^{k+l} = p_1 p_2 + (n_1 x + c_1 x) p_2 + (n_2 y + c_2 y) p_1 + n_1 n_2 xy$$
$$+ (n_2 c_1 + n_1 c_2 + c_1 c_2) xy \in P_0 + (xy).$$

So $P_0 + (xy)$ does not belong to the set E; so $xy \notin P_0$.

Thus P_0 is a prime ideal including I which does not contain b. So b does not lie in the radical of I.

This completes the proof.

COROLLARY. *For every ideal I of R we have* Rad (Rad I) = Rad I.

§ 14. Quotient rings

Let R be a commutative ring with identity. A subset S of R is called a *multiplicative system* if (1) the identity element e belongs to S and (2) S is closed under the multiplication operation of the ring, i.e. if a and b are elements of S then so is ab.

Example 1. For every element a of R the set of powers of a is a multiplicative system in R.

Example 2. If P is a prime ideal of R then the set of elements of R which do not belong to P is a multiplicative system in R.

Example 3. The set of elements of R which are not divisors of zero is a multiplicative system in R. In particular, if R is an integral domain, the set of non-zero elements of R is a multiplicative system in R.

If S is any subset of the ring R the set E of multiplicative systems of R which include S is non-empty, since R itself belongs to E. The intersection $\bar{S}$ of all these multiplicative systems is clearly also a multiplicative system including S; it is the smallest such system, and we call it the multiplicative system *generated by* S. It is easy to verify that $\bar{S}$ consists of all products of finite families of elements of S.

We consider now the product set $R \times \bar{S}$ and define a relation ρ in this set by writing $(a, s) \, \rho \, (a', s')$ if and only if there is an element t of $\bar{S}$ such that $t(as' - a's) = 0$. This relation is plainly reflexive and symmetric; we shall show that it is transitive, and hence an equivalence relation. So suppose $(a, s) \, \rho \, (a', s')$ and $(a', s') \, \rho \, (a'', s'')$. Then there exist elements t, t' of $\bar{S}$ such that $t(as' - a's) = t'(a''s' - a's'') = 0$; we have

$$0 = t's''t(as' - a's) - tst'(a''s' - a's'') = tt's'(as'' - a''s)$$

and so, since $tt's' \in \bar{S}$, it follows that $(a, s) \, \rho \, (a'', s'')$.

We form the quotient set of $R \times \bar{S}$ with respect to the equivalence relation ρ and denote it by $R[S^{-1}]$. Our next move is to define addition and multiplication operations in this set in such a way that it is turned into a ring. So let C_1, C_2 be any two equivalence classes in $R[S^{-1}]$; choose ordered pairs $(a_1, s_1), (a_2, s_2)$ from C_1, C_2 respectively. Then define $C_1 + C_2$ and $C_1 C_2$ to be the equivalence classes containing the ordered pairs $(a_1 s_2 + a_2 s_1, s_1 s_2)$ and $(a_1 a_2, s_1 s_2)$ respectively (these pairs actually belong to $R \times \bar{S}$ since s_1 and s_2 are in $\bar{S}$ and $\bar{S}$ is a multiplicative system). These classes depend only on C_1 and C_2, not on the choices of representatives $(a_1, s_1), (a_2, s_2)$. For suppose $(a_1', s_1'), (a_2', s_2')$ are alternative representatives for C_1, C_2; then there exists elements t_1, t_2 in $\bar{S}$ such that $t_1(a_1 s_1' - a_1' s_1) = 0$ and $t_2(a_2 s_2' - a_2' s_2) = 0$. It follows that

$$t_1 t_2 ((a_1 s_2 + a_2 s_1) s_1' s_2' - (a_1' s_2' + a_2' s_1') s_1 s_2) = 0$$

and

$$t_1 t_2 (a_1 a_2 s_1' s_2' - a_1' a_2' s_1 s_2) = 0.$$

That is to say

$$(a_1 s_2 + a_2 s_1, s_1 s_2) \, \rho \, (a_1' s_2' + a_2' s_1', s_1' s_2')$$

and

$$(a_1 a_2, s_1 s_2) \, \rho \, (a_1' a_2', s_1' s_2').$$

It is now easily verified that $R[S^{-1}]$ forms a commutative ring under these operations. The zero element of $R[S^{-1}]$ is $\eta(0, e)$ and the identity element is $\eta(e, e)$, where η denotes as usual the canonical surjection of $R \times \bar{S}$ onto the quotient set $R[S^{-1}]$. We define a mapping κ from R to $R[S^{-1}]$ by setting $\kappa(a) = \eta(a, e)$; it is easy to check that κ is a homomorphism. We notice also that for every element s of S the image $\kappa(s) = \eta(s, e)$ has a multiplicative inverse in

$R[S^{-1}]$, namely the equivalence class $\eta(e, s)$. From now on, we shall denote the equivalence class $\eta(a, s)$ of the element (a, s) of $R \times \bar{S}$ by the fraction notation a/s; we must observe then that $a/s = a'/s'$ if and only if $(a, s) \rho (a', s')$, i.e. if and only if there is an element t of $\bar{S}$ such that $t(as' - a's) = 0$.

The ring $R[S^{-1}]$ which we have just constructed is called the *quotient ring* or *ring of fractions* of R *defined by* S; the mapping κ is called the *canonical homomorphism* from R to $R[S^{-1}]$.

In general the canonical homomorphism κ is not a monomorphism. It is easy, however, to give conditions under which κ is a monomorphism.

THEOREM 14.1. *Let S be a subset of a commutative ring R, κ the canonical homomorphism from R to $R[S^{-1}]$. Then the kernel of κ consists of all elements a of R for which there exists an element t of $\bar{S}$ such that $ta = 0$.*

Proof. Let a be an element of R. Then $a \in \text{Ker } \kappa$ if and only if $\kappa(a) = \eta(a, e)$ is the zero element $\eta(0, e)$ of $R[S^{-1}]$, i.e. if and only if $(a, e) \rho (0, e)$. Hence, as asserted, $a \in \text{Ker } \kappa$ if and only if there is an element t of $\bar{S}$ such that $t(ae - 0e) = ta = 0$.

COROLLARY. *The canonical homomorphism κ is a monomorphism if and only if S contains only non-zero elements which are not divisors of zero.*

If R is a commutative ring and R_0 is the set of all elements of R which are not divisors of zero, the ring $R[R_0^{-1}]$ is called the *full ring of fractions* of R. In this case, of course, the canonical homomorphism κ is a monomorphism and we frequently identify R with its image under κ.

Suppose in particular that R is an integral domain. Here R_0 is the set of non-zero elements of R; so if a/b is a non-zero element of $R[R_0^{-1}]$, b/a is also an element of $R[R_0^{-1}]$ and is clearly an inverse for a/b. Thus in this case $R[R_0^{-1}]$ is a field. We call it the *field of fractions of* R.

Let R be a commutative ring with identity, S a subset of R; a homomorphism φ from R to a commutative ring T with identity is called *S-inverting* if (1) $\varphi(e_R) = e_T$, where e_R and e_T are the identity elements of R and T respectively and (2) every non-zero element of $\varphi(S)$ is invertible in T. Clearly the canonical homomorphism κ from

R to $R[S^{-1}]$ is S-inverting. A pair (R', κ') consisting of a commutative ring R' with identity and an S-inverting homomorphism κ' from R into R' is said to be *universal* for S-inverting homomorphisms from R if for every S-inverting homomorphism φ from R to a commutative ring T with identity there exists a unique homomorphism φ' from R' to T such that the diagram

is commutative.

THEOREM 14.2. *If R is a commutative ring with identity, S a subset of R, then the pair $(R[S^{-1}], \kappa)$ consisting of the ring of fractions of R defined by S and the canonical homomorphism from R to $R[S^{-1}]$ is universal for S-inverting homomorphisms from R.*

Proof. We have already remarked that $R[S^{-1}]$ is a commutative ring with identity and that κ is an S-inverting homomorphism from R to $R[S^{-1}]$.

Let φ be an S-inverting homomorphism from R to a commutative ring T with identity. As usual let $\bar{S}$ be the multiplicative system generated by S. Every element of $\bar{S}$ is a finite product of elements of S; so every element of $\varphi(\bar{S})$ is a finite product of elements of $\varphi(S)$ and hence is invertible in T.

We define a mapping φ' from $R[S^{-1}]$ to T by setting

$$\varphi'(a/s) = \varphi(a)\,(\varphi(s))^{-1}$$

for all elements a in R and s in $\bar{S}$. We have to verify, of course, that if $a'/s' = a/s$ then $\varphi(a')\,(\varphi(s'))^{-1} = \varphi(a)\,(\varphi(s))^{-1}$. But if $a'/s' = a/s$ there is an element t of $\bar{S}$ such that $t(a's - as') = 0$ and hence, since φ is a homomorphism,

$$\varphi(t)\,(\varphi(a')\varphi(s) - \varphi(a)\varphi(s')) = 0.$$

Since $s, s' \in S$ and $t \in \bar{S}$, the elements $\varphi(s), \varphi(s')$ and $\varphi(t)$ are all invertible in T and we deduce that $\varphi(a')\,(\varphi(s'))^{-1} = \varphi(a)\,(\varphi(s))^{-1}$, as required.

It is easily verified that φ' is a homomorphism from $R[S^{-1}]$ to T. Further, if a is any element of R we have $\varphi'\kappa(a) = \varphi'(a/e) = \varphi(a)(\varphi(e))^{-1} = \varphi(a)$; thus $\varphi'\kappa = \varphi$.

To show that φ' is the only homomorphism from $R[S^{-1}]$ to T with this property, suppose φ'' is another, i.e. suppose φ'' is a homomorphism from $R[S^{-1}]$ to T such that $\varphi''\kappa = \varphi$. Then for each element s of $\bar{S}$ we have

$$\varphi(s)\varphi''(e_R/s) = \varphi''(s/e_R)\varphi''(e_R/s) = \varphi(e_R) = e_T$$

where e_R and e_T are the identity elements of R and T respectively. Hence $\varphi''(e_R/s)$ is the inverse of $\varphi(s)$ in T. It follows that for every element a/s of $R[S^{-1}]$ we have

$$\varphi''(a/s) = \varphi''(a/e_R)\varphi''(e_R/s) = \varphi(a)(\varphi(s))^{-1} = \varphi'(a/s).$$

Thus $\varphi'' = \varphi'$ as required.

So $(R[S^{-1}], \kappa)$ is universal for S-inverting homomorphisms from R.

We now proceed to investigate the connexion between the ideals of R and those of the quotient ring of R defined by a multiplicative system S. First we consider a more general situation, in which we have two rings R and R_1 and a homomorphism α from R to R_1; we shall specialise this to the case where $R_1 = R[S^{-1}]$ and α is the canonical homomorphism κ from R to $R[S^{-1}]$. Let I be any ideal in R; then, if α is not an epimorphism, $\alpha(I)$ is not necessarily an ideal of R_1, but it certainly generates an ideal of R_1, which we call the *extension* of I to R_1 and denote by $I^{\sharp}$. An ideal I_1 of R_1 is called an *extended ideal* if there is an ideal I of R such that $I_1 = I^{\sharp}$. On the other hand, if J_1 is an ideal of R_1, the inverse image $\alpha^{-1}(J_1)$ is an ideal of R; we call this the *contraction* of J_1 to R and denote it by $J_1^{\flat}$. An ideal J of R is called a *contracted ideal*, if there is an ideal J_1 of R_1 such that $J_1^{\flat} = J$.

It follows easily from the definitions that, if I and J are ideals of R such that $I \subseteq J$, then $I^{\sharp} \subseteq J^{\sharp}$; and that if I_1 and J_1 are ideals of R_1 such that $I_1 \subseteq J_1$ then $I_1^{\flat} \subseteq J_1^{\flat}$. We contend also that if I is any ideal of R then $(I^{\sharp})^{\flat} \supseteq I$ and that if J_1 is any ideal of R_1 then $(J_1^{\flat})^{\sharp} \subseteq J_1$. To establish the first of these we have only to remark that if a is any element of I then $a \in \alpha^{-1}(\alpha(a)) \subseteq \alpha^{-1}(\alpha(I)) \subseteq \alpha^{-1}(I^{\sharp}) = (I^{\sharp})^{\flat}$. For the second, we notice that $\alpha(\alpha^{-1}(J_1)) \subseteq J_1$, i.e. $\alpha(J_1^{\flat}) \subseteq J_1$, from which it follows that $(J_1^{\flat})^{\sharp} \subseteq J_1$.

Let $\mathscr{E}$ be the set of extended ideals in R_1 and $\mathscr{C}$ the set of contracted ideals in R; both these sets are ordered by the inclusion relation. According to the previous paragraph the mappings φ and φ_1 from $\mathscr{C}$ to $\mathscr{E}$ and $\mathscr{E}$ to $\mathscr{C}$ respectively given by $\varphi(I) = I^\sharp$ for every contracted ideal I and $\varphi_1(I_1) = I_1^\flat$ for every extended ideal I_1 are increasing mappings. We claim that they are actually ordered set isomorphisms (relative to the inclusion relations in $\mathscr{C}$ and $\mathscr{E}$). So let I be any contracted ideal in R; then there is an ideal I_1 in R_1 such that $I = I_1^\flat$. Hence we have $\varphi_1\varphi(I) = \varphi_1\varphi(I_1^\flat) = ((I_1^\flat)^\sharp)^\flat$; so $\varphi_1\varphi(I) \supseteq I_1^\flat = I$. But since $(I_1^\flat)^\sharp \subseteq I_1$, we also have $((I_1^\flat)^\sharp)^\flat \subseteq I_1^\flat = I$. Thus $\varphi_1\varphi(I) = I$ for each ideal I in $\mathscr{C}$; so $\varphi_1\varphi$ is the identity mapping of $\mathscr{C}$. A similar argument shows that $\varphi\varphi_1$ is the identity mapping of $\mathscr{E}$. Thus φ and φ_1 are bijections, and hence isomorphisms as asserted.

Next let us consider the special case where R is a commutative ring with identity, $R_1 = R[S^{-1}]$ is the quotient ring of R defined by a multiplicative system S and α is the canonical homomorphism κ from R to R_1. An ideal I of R is said to be *prime to S* if for every element s of S the only elements x of R such that $sx \in I$ are the elements of I itself. We can now characterize the extended and contracted ideals and we obtain the following results.

THEOREM 14.3. *Let R be a commutative ring with identity, S a multiplicative system in R. Then there is an isomorphism (relative to the inclusion relation) from the set of all ideals of $R[S^{-1}]$ onto the set of ideals of R which are prime to S.*

Proof. First we show that every ideal of $R_1 = R[S^{-1}]$ is an extended ideal. So let J_1 be any ideal of R_1; we have already noticed that $(J_1^\flat)^\sharp \subseteq J_1$. To prove the reverse inclusion, let x_1 be any element of J_1; then there are elements a of R and s of S such that $x_1 = \kappa(a)(\kappa(s))^{-1}$. It follows that $\kappa(a) = \kappa(s)x_1$ lies in J_1 and hence that $a \in J_1^\flat$; so $\kappa(a) \in (J_1^\flat)^\sharp$ and hence finally $x_1 = \kappa(a)(\kappa(s))^{-1} \in (J_1^\flat)^\sharp$. Thus $J_1 = (J_1^\flat)^\sharp$ and so is an extended ideal.

Next we show that the contracted ideals of R are precisely those which are prime to S.

Suppose first that I is a contracted ideal of R. Then there is an ideal I_1 of R_1 such that $I = I_1^\flat$ and we have $(I^\sharp)^\flat = ((I_1^\flat)^\sharp)^\flat = I_1^\flat = I$. We shall deduce that I is prime to S. So let x be any element of R, s an element of S such that xs belongs to I; we claim that $x \in I$. Since

$a = xs \in I$ we have $\kappa(x)\kappa(s) = \kappa(a) \in \kappa(I) \subseteq I^\#$; hence $\kappa(x) = \kappa(a)\kappa(s)^{-1} \in I^\#$ and so $x \in \kappa^{-1}(I^\#) = (I^\#)^\flat = I$, as required.

Suppose conversely that I is prime to S. We shall show that $I = (I^\#)^\flat$, so that I is a contracted ideal; since we always have $I \subseteq (I^\#)^\flat$ we have only to prove the reverse inclusion. So let x be any element of $(I^\#)^\flat$; then $\kappa(x) \in I^\#$, the ideal of R_1 generated by $\kappa(I)$. Referring to §3 we see that

$$\kappa(x) = \sum_{i=1}^{k} y_i \kappa(a_i),$$

where (y_i), (a_i) are families of elements of R_1, I respectively. Each element y_i has the form $y_i = \kappa(x_i)\,(\kappa(s_i))^{-1}$ where $x_i \in R$ and $s_i \in S$. If we write $s = \prod_{i=1}^{k} s_i$ and $z_i = x_i s_1 \ldots \hat{s}_i \ldots s_k$ (the circumflex denoting the omission of the term over which it stands) we have $y_i = \kappa(z_i)\,(\kappa(s))^{-1}$ for $i = 1, \ldots, k$. Hence

$$\kappa(x) = \sum_{i=1}^{k} \kappa(z_i)\kappa(a_i)\,(\kappa(s))^{-1}$$

$$= \kappa\left(\sum_{i=1}^{k} z_1 a_i\right)(\kappa(s))^{-1}$$

$$= \kappa(a)\,(\kappa(s))^{-1}$$

where $a = \sum_{i=1}^{k} z_i a_i$ and hence $a \in I$. (Notice that this manoeuvre is just the familiar one of 'putting the fractions x_i/s_i over a common denominator s'.) It follows that $\kappa(sx - a) = 0$. Theorem 14.1 shows that there exists an element s' of S such that $s'(sx - a) = 0$, i.e. such that $(s's)x = s'a$, which is an element of I. Since $s's \in S$ and I is prime to S, we deduce that $x \in I$, as required. So $I = (I^\#)^\flat$ is a contracted ideal.

Since in general there is an isomorphism from the set of extended ideals onto the set of contracted ideals, it follows in the present special case that there is an isomorphism from the set of all ideals of $R[S^{-1}]$ onto the set of ideals of R which are prime to S.

COROLLARY. *If R is an Artinian or Noetherian ring, then so is $R[S^{-1}]$.*

Proof. Suppose R is an Artinian ring and let $(J_k)_{k \in \mathbb{N}}$ be a descending chain of ideals of $R[S^{-1}]$. Then $(J_k^\flat)_{k \in \mathbb{N}}$ is a descending chain of ideals of R which levels off, since R is Artinian. Hence the chain $(J_k^{\flat\#}) = (J_k)$ levels off, and so $R[S^{-1}]$ is Artinian.

The proof for the Noetherian case is similar.

We now restrict our attention to the prime ideals of $R[S^{-1}]$ and obtain the following result.

THEOREM 14.4. *Let R be a commutative ring with identity, S a multiplicative system in R. Then there is an isomorphism from the set of all proper prime ideals of $R[S^{-1}]$ onto the set of proper prime ideals of R which are disjoint from S.*

Proof. We show first that if P is a proper prime ideal of R then P is prime to S if and only if it is disjoint from S.

Let P be a prime ideal of R disjoint from S. Let s and x be elements of S and R respectively such that $sx \in P$. Since P is a prime ideal it follows that either $s \in P$ or $x \in P$; but since S and P are disjoint we cannot have $s \in P$. So $x \in P$, and this shows that P is prime to S.

On the other hand, let P be a proper prime ideal of R which is not disjoint from S. Let s be any element of $P \cap S$; then for every element x of R we have $sx \in P$, and hence P is not prime to S.

Suppose now P_1 is a proper prime ideal of $R_1 = R[S^{-1}]$; we claim that $P_1^\flat$ is a proper prime ideal of R. If $P_1^\flat$ is not proper, we have $P_1^\flat = R$, whence $R_1 = R^\sharp = (P_1^\flat)^\sharp \subseteq P_1$, which contradicts the hypothesis that P_1 is proper. Hence $P_1^\flat$ is proper. To show that $P_1^\flat$ is prime, suppose a and b are elements of R such that $ab \in P_1^\flat$; then $\kappa(a)\kappa(b) = \kappa(ab) \in P_1$ and so either $\kappa(a)$ or $\kappa(b)$ belongs to P_1, i.e. either a or b belongs to $\kappa^{-1}(P_1) = P_1^\flat$, which is therefore a prime ideal. Since $P_1^\flat$ is a contracted ideal it is prime to S and hence, according to the preceding paragraph, disjoint from S.

Conversely let P be a prime ideal of R which is disjoint from S and hence prime to S; thus P is a contracted ideal. We shall show that $P^\sharp$ is a proper prime ideal of R_1. As in the proof of Theorem 14.3, we can prove that every element of $P^\sharp$ can be expressed in the form $\kappa(p)(\kappa(s))^{-1}$ where $p \in P$ and $s \in S$. Thus, if $P^\sharp = R_1$ the identity element of R_1 belongs to $P^\sharp$ and so there are elements p of P and s of S such that $\kappa(p) = \kappa(s)$, whence $s - p \in \operatorname{Ker} \kappa$; according to Theorem 14.1, there is an element s' of S such that $s'(s - p) = 0$, i.e. $s's = s'p$. But $s's \in S$ and $s'p \in P$; so P and S are not disjoint. This contradiction shows that $P^\sharp$ is a proper ideal of R_1. To show that $P^\sharp$ is a prime ideal, let $x_1 = \kappa(a)(\kappa(s))^{-1}$ and $y_1 = \kappa(b)(\kappa(t))^{-1}$ be elements of R_1 such that $x_1 y_1 \in P^\sharp$; then there are elements p' of P and s' of S such that $x_1 y_1 = \kappa(p')(\kappa(s'))^{-1}$. We deduce that

$\kappa(a)\kappa(b)\kappa(s') = \kappa(s)\kappa(t)\kappa(p')$ and hence that $abs' - stp' \in \text{Ker } \kappa$. Thus there is an element t' of S such that $(abs' - stp')t' = 0$; then $ab(s't') = stt'p \in P$. Since P is a prime ideal it follows that either $a \in P$ or $b \in P$ or $s't' \in P$. But the last possibility is ruled out since $s't' \in S$ and P is disjoint from S; so either $a \in P$ or $b \in P$, say $a \in P$. Then $\kappa(a) \in \kappa(P) \subseteq P^\#$ and so $x_1 = \kappa(a)\,(\kappa(s))^{-1} \in P^\#$; consequently $P^\#$ is a prime ideal of R_1 as required.

It follows from all this that the mapping φ from $\mathscr{C}$ (the set of contracted ideals of R) to $\mathscr{E}$ (the set of extended ideals of R_1) given by $(I) = I^\#$ for every contracted ideal I maps the set $\mathscr{P}$ of proper prime ideals of R disjoint from S into the set $\mathscr{P}_1$ of proper prime ideals of R_1 and that the mapping φ_1 from $\mathscr{E}$ to $\mathscr{C}$ maps $\mathscr{P}_1$ into $\mathscr{P}$. Hence, since φ and φ_1 are inverse isomorphisms, their restrictions to $\mathscr{P}$ and $\mathscr{P}_1$ respectively are inverse isomorphisms also.

This completes the proof.

§ 15. Integral Dependence

In this section all the rings we shall consider will be integral domains with identity elements.

Let R and R' be two such rings such that $R \subseteq R'$. We remark first that the identity elements of R and R' coincide; for if a is any non-zero element of R we have $e_R a = a = e_{R'} a$ and hence $e_R = e_{R'}$ since R has no divisors of zero.

Now let x be any element of R'; we say that x is *integral over R* if there is a natural number $n \geq 1$ and a family $(a_i)_{i \in [1, n]}$ of elements of R such that

$$x^n + a_1 x^{n-1} + a_2 x^{n-2} + \ldots + a_n = 0. \qquad [15.1]$$

Of course if a is any element of R, then a is integral over R—we simply take $n = 1$ and $a_1 = -a$. If every element of R' is integral over R we say that the ring R' is *integral over R* or *integrally dependent* on R.

We may regard R' as a left R-module under the scalar multiplication λ defined by setting $\lambda(a, x) = ax$ for all elements a of R and x of R', the product on the right being formed according to the ring multiplication operation in R'. If V is any submodule of R' considered as a left R-module in this way and x is any element of R' we shall denote by xV the set of all elements of R' of the form xv where $v \in V$; it is easily verified, using the commutativity of R', that xV is also a left R-module. We use these ideas to give an equivalent condition for integral dependence.

THEOREM 15.1. *Let R and R' be integral domains with identity such that R is a subring of R'. An element x of R' is integral over R if and only if there exists a finitely generated non-zero submodule V of R' (considered as left R-module) such that xV is included in V.*

Proof. (1) Suppose x is integral over R, satisfying the condition [15.1]. Let V be the R-submodule of R' generated by the set $\{e, x, \ldots, x^{n-1}\}$. Each element v of V has the form $v = r_0 + r_1 x + \ldots + r_{n-1}x^{n-1}$ where $r_0, r_1, \ldots, r_{n-1}$ are elements of R. Then we have

$$xv = r_0 x + r_1 x^2 + \ldots + r_{n-2}x^{n-1} - r_{n-1}(a_1 x^{n-1} + \ldots + a_n);$$

so $xV \subseteq V$.

(2) Conversely, let V be a finitely generated non-zero submodule of R' such that $xV \subseteq V$. Let $\{v_1, \ldots, v_n\}$ be a generating system for V; we have $n \geqslant 1$ since V is non-zero. Since $xV \subseteq V$, it follows that for $i = 1, \ldots, n$ we have $xv_i \in V$; hence there is an $n \times n$ matrix $A = (a_{ij})$ with elements in R such that

$$xv_i = a_{i1}v_1 + a_{i2}v_2 + \ldots + a_{in}v_n \ (i = 1, \ldots, n).$$

We may consolidate these equations into a single matrix equation

$$(xI - A)\mathbf{v} = \mathbf{0}$$

where I is the $n \times n$ identity matrix and $\mathbf{v}$ and $\mathbf{0}$ are the column

vectors $\begin{bmatrix} v_1 \\ v_2 \\ \cdot \\ \cdot \\ v_n \end{bmatrix}$ and $\begin{bmatrix} 0 \\ 0 \\ \cdot \\ \cdot \\ 0 \end{bmatrix}$ respectively. Premultiplying by the adjugate

(adjoint) of the matrix $xI - A$ we obtain

$$\det (xI - A)\,\mathbf{v} = 0$$

(see A. C. Aitken, *Determinants and matrices*, Chapter III). Since $\mathbf{v}$ is not the zero vector, it follows that $\det (xI - A) = 0$, and on expanding the determinant, we obtain a relation of the form [15.1]. So x is integral over R.

We prove next a result which is in effect a generalisation of the first part of the preceding theorem.

THEOREM 15.2. *Let R and R' be integral domains with identity such that R is a subring of R'. If $x_1, \ldots, x_m$ are elements of R' all of which are integral over R, the subring $R[x_1, \ldots, x_m]$ is a finitely generated submodule of the R-module R'.*

Proof. We proceed by induction on m.

If $m = 1$ and $x_1 = x$ satisfies the relation [15.1], it follows that $R[x]$ is included in the R-module V generated by $\{e, x, \ldots, x^{n-1}\}$; but of course $V \subseteq R[x]$. Thus $R[x] = V$.

Suppose now we have established for some natural number k that $R_k = R[x_1, \ldots, x_k]$ is a finitely generated R-submodule of R', generated say by $\{v_1, \ldots, v_r\}$. Since x_{k+1} is integral over R, it is certainly integral over R_k and hence $R_{k+1} = R[x_1, \ldots, x_{k+1}] = R_k[x_{k+1}]$ is a finitely generated R_k-submodule of R' (using the result of the previous paragraph). If $\{w_1, \ldots, w_s\}$ is a set of generators for R_{k+1} as R_k-module, it follows that $\{v_1 w_1, \ldots, v_r w_s\}$ is a set of generators for R_{k+1} as an R-module. This completes the induction.

COROLLARY. *Let R and R' be integral domains with identity elements such that R is a subring of R'. The subset of R' consisting of all elements of R' which are integral over R is a subring of R' which includes R.*

Proof. We have already remarked that every element of R is integral over R.

Let x and y be elements of R' which are integral over R. According to the Theorem $R[x, y]$ is a finitely generated submodule of the left R-module R'; write $V = R[x, y]$. Since $x - y$ and xy belong to $R[x, y]$, it follows that $(x - y)V$ and xyV are included in V. Theorem 15.1 now shows that $x - y$ and xy are integral over R. The required result now follows from Theorem 3.2.

The subring of R' consisting of elements which are integral over R is called the *relative integral closure* of R in R'; if this relative integral closure just consists of the elements of R itself we say that R is *relatively integrally closed* in R'. If $R' = F$, the field of fractions of the integral domain R, and we identify R with its image under the canonical monomorphism κ, the relative integral closure of R in F is called simply the *integral closure* of R; if this integral closure is the ring R itself, then R is said to be *integrally closed*.

THEOREM 15.3. *Let R, R′, R″ be integral domains with identity such that R is a subring of R′ and R′ is a subring of R″. If R′ is integral over R and R″ is integral over R′ then R″ is integral over R.*

Proof. Let x be any element of R''. Since x is integral over R' there are elements $b_1, \ldots, b_k$ of R' such that

$$x^k + b_1 x^{k-1} + \ldots + b_k = 0.$$

Let $B = R[b_1, \ldots, b_k]$; since $b_1, \ldots, b_k$ are integral over R, it follows from Theorem 15.2 that B is a finitely-generated R-module, generated over R by $\{x_1, \ldots, x_m\}$ say. The element x is of course integral over B; so, by Theorem 15.2 again, $B[x]$ is a finitely generated B-module, generated over B by $\{y_1, \ldots, y_n\}$ say. Then $B[x]$ is generated as an R-module by the set of products $\{x_i y_j\}$ and hence is a finitely generated R-module. Since $xB[x] \subseteq B[x]$, it follows from Theorem 15.1 that x is integral over R, as required.

COROLLARY. *Let R and R′ be integral domains with identity such that R is a subring of R′. Then the relative integral closure of R in R′ is relatively integrally closed in R′. In particular the integral closure of R is integrally closed.*

Example 1. Let R be a unique factorisation domain. We claim that R is integrally closed. So let x be any element of the field of fractions F of R which is integral over R; say $x = a/b$ where $a, b \in R$, and we may assume without loss of generality that a and b are relatively prime, i.e. that the ' fraction ' a/b is ' in its lowest terms '. Then there exist elements $a_1, \ldots, a_n$ of R such that

$$x^n + a_1 x^{n-1} + \ldots + a_n = 0$$

whence

$$a^n + a_1 b a^{n-1} + \ldots + a_n b^n = 0$$

and so

$$a^n = -(a_1 a^{n-1} + \ldots + a_n b^{n-1}) b.$$

It follows at once that b is a unit of R; for if not, b would have an irreducible factor, p say. According to the uniqueness of factorisation this irreducible element p would be a factor of a^n, and hence (since irreducible elements in a unique factorisation domain are prime) p

would divide a. This, however, contradicts the assumption that a and b are relatively prime. Hence b is a unit of R, and so $x = a/b \in R$. Thus R is integrally closed in F.

Example 2. Let R be an integral domain with identity, F its field of fractions, and S a multiplicative system in R. By means of obvious identifications, we may consider R as a subring of $R[S^{-1}]$, $R[S^{-1}]$ as a subring of F, and F as the field of fractions of $R[S^{-1}]$. We shall show that, if R is integrally closed, so is $R[S^{-1}]$. So suppose x is an element of F integral over $R[S^{-1}]$; then there exist elements $b_1, \ldots, b_n$ of $R[S^{-1}]$ such that

$$x^n + b_1 x^{n-1} + \ldots + b_n = 0. \qquad [15.2]$$

We may write $b_i = a_i/s_i$ where $a_i \in R$ and $s_i \in S$ $(i = 1, \ldots, n)$. Set $s = s_1 s_2 \ldots s_n$ and $s_i' = s_1 \ldots \hat{s}_i \ldots s_n$ $(i = 1, \ldots, n)$, where the circumflex as usual denotes the omission of the term over which it stands. Then, multiplying $[15.2]$ by s^n, we obtain

$$(sx)^n + a_1 s_1'(sx)^{n-1} + a_2 s_2' s(sx)^{n-2} + \ldots + a_n s_n' s^{n-1} = 0,$$

which shows that $z = sx$ is integral over R. Since R is integrally closed, it follows that $z \in R$; since S is a multiplicative system, $s \in S$; hence $x = z/s \in R[S^{-1}]$ as required.

§ 16. Dedekind Domains

The unique factorisation properties of the ring of integers and of rings of polynomials with coefficients in a field were so familiar that it came as a shock to nineteenth-century mathematicians to discover that there exist integral domains which are not unique factorisation domains. But this discovery, however shocking, was a very fruitful one for mathematics since it led to the development by Kummer and Dedekind of the theory of ideals, a theory which not only provided a new and more extensive concept of unique factorisation but also had far-reaching consequences for the study of rings.

Before discussing the modification of the notion of unique factorisation which we have just referred to, we shall describe the classical example of an integral domain which is not a unique factorisation domain as defined in §12.

Let R be the set of complex numbers of the form $a + b\sqrt{5i}$, where a and b are integers. It is easily verified that R is a subring of the complex number field $\mathbf{C}$; it is thus an integral domain, and it has an identity element $e = 1 + 0\sqrt{5i}$. We define the mapping v from R to the set of natural numbers $\mathbf{N}$ by setting $v(x) = a^2 + 5b^2$ for every element $x = a + b\sqrt{5i}$ of R; it is quickly verified by direct computation that for all pairs of elements x_1, x_2 of R we have $v(x_1 x_2) = v(x_1)v(x_2)$.

If $x = a + b\sqrt{5i}$ and $v(x) = a^2 + 5b^2 = 1$, it follows that $a = \pm 1$, $b = 0$; so x is either e or $- e$, and hence is a unit of R. Conversely, if u is a unit of R, with inverse v, we have $v(u)v(v) = v(uv) = v(e) = 1$; hence $v(u) = 1$ and so u is either e or $- e$.

We claim that the elements 2, 3, $1 + \sqrt{5i}$ and $1 - \sqrt{5i}$ are irreducible elements of R. Suppose, for example, that $1 + \sqrt{5i} = xy$ where x and y are both non-units of R. Then we have

$$v(x) \neq 1, \; v(y) \neq 1, \; v(x)v(y) = v(xy) = 6; \qquad [16.1]$$

hence either $v(x) = 3$ and $v(y) = 2$ or else $v(x) = 2$ and $v(y) = 3$. But it is impossible to find integers a and b such that $v(a + b\sqrt{5i}) = a^2 + 5b^2 = 2$ or 3; hence it is impossible to find elements x and y of R which satisfy [16.1]. Thus $1 + \sqrt{5i}$ is irreducible. Similar arguments show that 2, 3 and $1 - \sqrt{5i}$ are irreducible. We now observe that

$$2.3 = 6 = (1 + \sqrt{5i})(1 - \sqrt{5i}). \qquad [16.2]$$

Since the only associates of 2 are 2 and -2 it follows that the element 6 of R has two essentially distinct factorisations into irreducible elements.

Let us try to find some guidance out of the unpleasant situation just revealed by considering again the case of a principal ideal domain. So let R be a principal ideal domain with identity; if a is any non-unit of R it follows from Theorem 12.5 that a has an essentially unique factorisation

$$a = p_1 p_2 \ldots p_r \qquad [16.3]$$

say, as a product of irreducible elements of R. From [16.3] we deduce easily that the principal ideal generated by a is the product of the principal ideals generated by the irreducible elements $p_1, \ldots, p_r$:

$$(a) = (p_1)(p_2)\ldots(p_r). \qquad [16.4]$$

The essential uniqueness of the factorisation [16.3] implies that the ideal factorisation [16.4] is also essentially unique, in the sense that if $(a) = (p_1')(p_2')\dots(p_s')$ is another such factorisation (with $p_1',\dots,p_s'$ irreducible) then $r = s$ and there is a bijection π from $[1, r]$ onto itself such that $(p_i) = (p_{\pi(i)}')$. According to Example 2 of §13, the principal ideals (p_i) are all prime ideals of R; Example 3 of §13 shows that all prime ideals of R are in fact maximal ideals.

This brief discussion perhaps suggests that instead of looking at the factorisation of elements it might be of interest to study the factorisation of ideals in an integral domain. An integral domain in which every non-zero proper ideal can be expressed as the product of a finite family of proper prime ideals is called a *Dedekind domain*. We remark that in this definition we have not said anything about the uniqueness of such prime ideal factorisation: this is not an oversight, for we shall prove shortly that the uniqueness is a consequence of the existence of a prime ideal factorisation.

Before proceeding with our study of the factorisation of ideals we introduce a useful piece of machinery. Let R be an integral domain, F its field of fractions. A subset A of F is called a *fractional ideal* of R if (1) A is an R-module and (2) there is a non-zero element d of R such that dA is included in R. Condition (1) requires that for all elements a_1, a_2 of A we have $a_1 + a_2 \in A$ and that for every element a of A and r of R we have $ra \in A$; the requirement of condition (2) is that there should exist an element d of R such that for every fraction r/s in A the product $d(r/s)$ belongs to R—in other words, that the fractions in A should 'have bounded denominators'. Clearly every ideal in the domain R is a fractional ideal (in condition (2) we need only take d to be the identity element e); for the present discussion these ordinary ideals of R will be referred to as *integral ideals*.

If A and B are fractional ideals of R we define the product AB to be the subset of F consisting of all elements which can be expressed in the form $\sum_{i \in I} a_i b_i$ where $(a_i)_{i \in I}$ and $(b_i)_{i \in I}$ are families of elements of A and B respectively, indexed by the same finite index set I. We contend that this product is also a fractional ideal of R. Condition (1) is clearly satisfied; to show that condition (2) is also satisfied, suppose c and d are elements of R such that $cA \subseteq R$ and $dB \subseteq R$; it then follows at once that $(cd)(AB) \subseteq R$. It is easily verified that if B_1 and B_2 are fractional ideals such that $B_1 \subseteq B_2$ then $AB_1 \subseteq AB_2$.

Let $FI(R)$ be the set of non-zero fractional ideals of R. The multiplication operation which we have just defined is an internal law of composition on $FI(R)$, for it is clear that if A and B are non-zero fractional ideals so is AB; we verify easily that this law of composition is both associative and commutative (cf. §3), and that the integral ideal R is an identity for the law of composition. The question which now obviously suggests itself is whether any (or possibly all) of the fractional ideals are invertible. To this end we introduce the notion of the reciprocal of a fractional ideal: if A is a fractional ideal of R its *reciprocal* is defined to be the subset A^* of F consisting of those elements x of F such that $xa \in R$ for every element a of A. This set A^* is clearly an R-module; and if d is any non-zero element of $A \cap R$ (which clearly does not consist of the zero element alone), we have $dA^* \subseteq R$; so A^* is a fractional ideal. It is easy to check that if A_1 and A_2 are fractional ideals such that $A_1 \subseteq A_2$ then $A_2^* \subseteq A_1^*$.

It follows at once from the definition that $A^*A \subseteq R$, but in general this is all we can say—in other words we cannot assert in general that $A^*A = R$. Suppose, however, that A has an inverse in $FI(R)$, i.e. that there is a fractional ideal A^{-1} such that $A^{-1}A = R$. Then certainly $A^{-1} \subseteq A^*$; but since $AA^* \subseteq R$ it follows that $A^{-1}(AA^*) \subseteq A^{-1}R$, whence $A^* = RA^* = (A^{-1}A)A^* \subseteq A^{-1}$. So we have shown that if a fractional ideal A of R has an inverse in $FI(R)$ then that inverse must be its reciprocal A^*.

If $x = a/b$ is a non-zero element of F ($a, b \in R$, $a, b \neq 0$) the set of elements of F of the form $rx = ra/b$ (where $r \in R$) is a fractional ideal of R, which we denote by (x). It is easy to check that (x) is invertible in $FI(R)$, its inverse being the fractional ideal (x^{-1}).

Next, let B be any non-zero fractional ideal of R; suppose $dB \subseteq R$. Then if b is any element of B there is an element b_1 of R such that $db = b_1$, whence $b = b_1/d$; thus every fraction in B can be written with d as denominator. Let B_1 be the subset of R consisting of the numerators of all such fractions, i.e. $b_1 \in B_1$ if and only if, $b_1/d \in B$. Then it is easy to verify that B_1 is an (integral) ideal of R; if we denote the integral ideal (d) by B_2, then B_2 is invertible and we have $B = B_1 B_2^{-1}$.

Let A be a non-zero proper integral ideal of an integral domain R. We say that A has an *essentially unique factorisation* into prime ideals if there exists a representation

$$A = P_1 P_2 \ldots P_r$$

of A as a product of proper (integral) prime ideals of R and if whenever

$$A = P_1 P_2 \ldots P_r = Q_1 Q_2 \ldots Q_s$$

are two such representations of A then $r = s$ and there is a bijection π from $[1, r]$ to itself such that $P_i = Q_{\pi(i)}$ $(i = 1, \ldots, r)$. We are now in a position to prove the result mentioned earlier—that if R is a Dedekind domain then every non-zero proper integral ideal of R has an essentially unique factorisation into prime ideals. The proof proceeds in two stages: first we establish a special case of the desired result; we then use this in the second stage to establish a result of independent interest, which allows us to reduce the general case to the special case already established.

THEOREM 16.1. *Let R be a Dedekind domain. Every non-zero proper prime ideal of R is a maximal ideal of R, and every non-zero proper integral ideal of R has an essentially unique factorisation into prime ideals.*

Proof. (1) First we prove that if a non-zero proper integral ideal A in any integral domain R has a factorisation into *invertible* proper prime ideals then this factorisation is essentially unique.

So suppose we have

$$A = P_1 P_2 \ldots P_r$$

where $P_1, \ldots, P_r$ are invertible proper prime ideals of R. Let

$$A = Q_1 Q_2 \ldots Q_s$$

be another factorisation of A into proper prime ideals of R.

We proceed by induction on r. If $r = 1$, then $A = P_1$ is a prime ideal. We shall prove that $s = 1$ also. So suppose $s > 1$; since $Q_1 \ldots Q_s = P_1$, it follows from Theorem 13.3. that at least one of the ideals $Q_1, \ldots, Q_s$ is included in P_1; say $Q_1 \subseteq P_1$. Since P_1 is invertible, by hypothesis, we have

$$\begin{aligned} Q_2 \ldots Q_s &= R Q_2 \ldots Q_s \\ &= P_1^{-1} P_1 Q_2 \ldots Q_s \subseteq P_1^{-1} Q_1 Q_2 \ldots Q_s = P_1^{-1} P_1 = R. \end{aligned}$$

This is a contradiction, since $Q_2, \ldots, Q_s$ are proper ideals and their product is included in each one of them. Hence $s = 1$, and A has an essentially unique factorisation.

Now we make the inductive hypothesis that every ideal which is expressible as a product of $r - 1$ invertible proper prime ideals has an essentially unique factorisation. Let P_m be a minimal element of the set $\{P_1, \ldots, P_r\}$. Since we have

$$Q_1 \ldots Q_s = P_1 \ldots P_r \subseteq P_m,$$

Theorem 13.3 shows that at least one of the ideals $Q_1, \ldots, Q_s$ is included in P_m, say $Q_n \subseteq P_m$. Similarly, since

$$P_1 \ldots P_r = Q_1 \ldots Q_s \subseteq Q_n$$

at least one of the ideals $P_1, \ldots, P_r$ is included in Q_n, say $P_k \subseteq Q_n$. Thus we have $P_k \subseteq Q_n \subseteq P_m$. Since P_m is minimal in the set $\{P_1, \ldots, P_r\}$ it follows that $P_k = Q_n = P_m$. Hence we have

$$P_m^{-1}A = P_1 \ldots \hat{P}_m \ldots P_r = Q_1 \ldots \hat{Q}_n \ldots Q_s$$

(where the circumflexes have the usual significance). Now $P_m^{-1}A$ is expressed as a product of $r - 1$ invertible prime ideals; hence, by the inductive hypothesis, $r - 1 = s - 1$ and there is a bijection π' from $\{1, \ldots, \hat{m}, \ldots, r\}$ onto $\{1, \ldots, \hat{n}, \ldots, r\}$ such that $P_i = Q_{\pi'(i)}$ ($i = 1, \ldots, m - 1, m + 1, \ldots, r$). It follows that $r = s$ and that we can define a bijection π from $[1, r]$ onto itself such that $P_i = Q_{\pi(i)}$ ($i = 1, \ldots, r$); namely, we set $\pi(i) = \pi'(i)$ for $i \neq m$ and $\pi(m) = n$.

This completes the induction.

(2) Next we show that every invertible proper prime ideal of a Dedekind domain R is maximal.

Let P be an invertible proper prime ideal of R and suppose that P is not a maximal ideal, i.e. that there exists an ideal A of R such that $P \subset A \subset R$. If a is any element of A which does not belong to P, the ideals $P + (a)$ and $P + (a^2)$ are included in A and hence are proper ideals of the Dedekind domain R. Consequently these ideals can be expressed as products of prime ideals, say

$$P + (a) = P_1 \ldots P_m \quad \text{and} \quad P + (a^2) = Q_1 \ldots Q_n. \quad [16.5]$$

Since P is a prime ideal the residue class ring R/P is an integral domain. Let η be the canonical epimorphism from R onto R/P; set $\bar{a} = \eta(a)$. Since a does not belong to P, $\bar{a}$ is a non-zero element of R/P and hence so is $\bar{a}^2$; thus the ideals $(\bar{a})$ and $(\bar{a}^2)$ are invertible in $FI(R/P)$. Applying the mapping η to the equations [16.5] and noticing that $\eta(P)$ is the zero ideal of R/P, we have

$$(\bar{a}) = \eta(P_1)\ldots\eta(P_m) \quad \text{and} \quad (\bar{a}^2) = \eta(Q_1)\ldots\eta(Q_n). \quad [16.6]$$

It follows easily from [16.5] that the ideals $P_1, \ldots, P_m, Q_1, \ldots, Q_n$ all include P. So, according to the Corollary of Theorem 13.2, the ideals $\eta(P_1), \ldots, \eta(P_m), \eta(Q_1), \ldots, \eta(Q_n)$ are proper prime ideals of R/P; and according to Theorem 1.10 they are all invertible.

We are now in a situation where we can apply the result of the first part of the proof. The second equation in [16.6] gives an expression of the ideal $(\bar{a}^2)$ as a product of invertible proper prime ideals of R/P; so (1) assures us that this factorisation is essentially unique. But if we square the first equation in [16.6], we obtain another factorisation of $(\bar{a}^2)$:

$$(\bar{a}^2) = (\bar{a})^2 = (\eta(P_1))^2 \ldots (\eta(P_m))^2.$$

Write $\bar{P}_{2i-1} = \bar{P}_{2i} = \eta(P_i) \, (i = 1, \ldots, m)$; then

$$(\bar{a})^2 = \bar{P}_1 \ldots \bar{P}_{2m}.$$

It follows that $n = 2m$ and that there is a bijection π from $[1, 2m]$ onto itself such that $\bar{P}_j = \eta(Q_{\pi(j)}) \, (j = 1, \ldots, n)$. Thus we have

$$\eta(Q_{\pi(2i-1)}) = \eta(Q_{\pi(2i)}) = \eta(P_i) \qquad (i = 1, \ldots, m).$$

We have already remarked that the ideals $P_1, \ldots, P_m, Q_1, \ldots, Q_n$ all include P; hence, using Theorem 4.5, we deduce that

$$Q_{\pi(2i-1)} = Q_{\pi(2i)} = P_i \quad (i = 1, \ldots, m)$$

and hence, substituting in [16.5] that $(P + (a))^2 = P + (a^2)$.

We now have

$$P \subseteq P + (a^2) = (P + (a))^2 \subseteq P^2 + (a).$$

Hence, if x is any element of P, there exist elements y in P^2 and r in R such that $x = y + ra$; then $ra = x - y \in P$ and so, since a does not belong to P, we have $r \in P$. This shows that $P \subseteq P^2 + P(a)$; but of course $P^2 + P(a) \subseteq P$. So $P = P(P + (a))$.

At this point we use the hypothesis that P is invertible; multiplying the last equation by P^{-1} we obtain $R = P + (a)$. This is a contradiction, and hence P is indeed a maximal ideal, as asserted.

(3) Now let P be any non-zero proper prime ideal of the Dedekind domain R. If x is any non-zero element of P the ideal (x) has a factorisation into proper prime ideals:

$$(x) = P_1 \ldots P_k.$$

Since (x) is invertible, it follows from Theorem 1.10 that the ideals $P_1, \ldots, P_k$ are all invertible and hence maximal (by part (2) above). Since

$$(x) = P_1 \ldots P_k \subseteq P$$

and P is a prime ideal, at least one of the ideals $P_1, \ldots, P_k$ is included in P, say $P_i \subseteq P$. We have just shown, however, that P_i is a maximal ideal of R; hence $P = P_i$ and so is maximal and invertible.

(4) Finally let A be any non-zero proper integral ideal in the Dedekind domain R. Then A can be expressed as a product of proper prime ideals. According to part (3) of the proof these prime ideals are all invertible. Hence it follows at once from part (1) that the factorisation of A is essentially unique.

This completes the proof.

COROLLARY. *Every non-zero integral ideal of a Dedekind domain is invertible.*

Let R be a Dedekind domain; let $\mathscr{P}$ be the set of non-zero proper prime ideals of R. If A is any non-zero proper integral ideal of R, it has a factorisation

$$A = P_1 \ldots P_k \qquad [16.7]$$

into non-zero proper prime ideals of R. For each prime ideal P in $\mathscr{P}$ let us set $v_P(A) = \text{Card} \{i \in [1, k] \mid P_i = P\} = $ the number of times the prime ideal P occurs in the factorisation $[16.7]$; it follows from the essential uniqueness of the factorisation of A that $v_P(A)$ depends only on A and not on the particular representation $[16.7]$. Clearly, $v_P(A)$ is non-zero for at most finitely many prime ideals P in $\mathscr{P}$, so it makes sense to write

$$A = \prod_{P \in \mathscr{P}} P^{v_P(A)} \qquad [16.8]$$

since $P^{v_P(A)} = P^0 = R$ for all but a finite number of prime ideals P, and these factors are irrelevant. The expression $[16.8]$ is clearly unique.

Next let B be any non-zero fractional ideal of R; suppose $B = B_1 B_2^{-1}$ where B_1, B_2 are integral ideals. Then we have unique expressions of the form $[16.8]$ for B_1 and B_2, say

$$B_1 = \prod_{P \in \mathscr{P}} P^{v_P(B_1)}, \quad B_2 = \prod_{P \in \mathscr{P}} P^{v_P(B_2)} \qquad [16.9]$$

from which we deduce the expression

$$B = \prod_{P \in \mathscr{P}} P^{v_P(B_1) - v_P(B_2)} \qquad [16.10]$$

for B. We claim that this expression is also unique. So suppose we also have $B = B_1'(B_2')^{-1}$ where B_1', B_2' are integral ideals; if B_1' and B_2' have factorisations corresponding to [16.9] we have another expression for B:

$$B = \prod_{P \in \mathscr{P}} P^{v_P(B_1') - v_P(B_2')} \qquad [16.11]$$

We must show that this is the same as [16.10]. Since $B_1(B_2)^{-1} = B_1'(B_2')^{-1}$ we have $B_1 B_2' = B_1' B_2$, whence

$$\prod_{P \in \mathscr{P}} P^{v_P(B_1) + v_P(B_2')} = \prod_{P \in \mathscr{P}} P^{v_P(B_1') + v_P(B_2)}.$$

From the uniqueness of such expressions for integral ideals we deduce that

$$v_P(B_1) + v_P(B_2') = v_P(B_1') + v_P(B_2)$$

for every prime ideal P in $\mathscr{P}$. It follows immediately that the expressions [16.10] and [16.11] are the same. If we write $v_P(B) = v_P(B_1) - v_P(B_2)$ we see that every non-zero fractional ideal B of R can be expressed uniquely in the form

$$B = \prod_{P \in \mathscr{P}} P^{v_P(B)}$$

where the indices $v_P(B)$ are integers (positive, negative or zero) such that $v_P(B) = 0$ for all but a finite number of prime ideals P in $\mathscr{P}$. It is clear that B is invertible, with inverse $B^{-1} = \prod_{P \in \mathscr{P}} P^{-v_P(B)}$. Clearly also, the fractional ideal B is integral if and only if $v_P(B) \geqslant 0$ for every prime ideal P in $\mathscr{P}$.

Let $B = \prod_{P \in \mathscr{P}} P^{v_P(B)}$, $C = \prod_{P \in \mathscr{P}} P^{v_P(C)}$ be fractional ideals of R. Then clearly B is included in C if and only if BC^{-1} is included in R (i.e. BC^{-1} is an integral ideal). Since $BC^{-1} = \prod_{P \in \mathscr{P}} P^{v_P(B) - v_P(C)}$ it follows that B is included in C if and only if $v_P(B) - v_P(C) \geqslant 0$ (i.e. $v_P(B) \geqslant v_P(C)$) for every prime ideal P in $\mathscr{P}$.

The next theorem gives two criteria for an integral domain with identity to be a Dedekind domain.

THEOREM 16.2. *Let R be an integral domain with identity. Then the* *following properties are equivalent*:

    (a) *R is a Dedekind domain*;

    (b) *R is a Noetherian domain, integrally closed in its field of fractions, and every non-zero proper prime ideal of R is maximal*;

    (c) *the semigroup FI(R) of non-zero fractional ideals of R is a group.*

Proof. (1) First we show that (a) implies (b). So suppose R is a Dedekind domain.

Let A be any (integral) ideal of R. By the Corollary to Theorem 16.1, A is invertible. Since the identity element $e \in R = A^{-1}A$, there are elements $a_1, \ldots, a_n$ of A and $a'_1, \ldots, a'_n$ of A^{-1} such that $e = a'_1 a_1 + \ldots + a'_n a_n$. We showed earlier (on page 123) that the inverse of an invertible ideal must be its reciprocal; so, if a is any element of A, all the elements $aa'_1, \ldots, aa'_n$ belong to R. But we have

$$a = ae = (aa'_1) a_1 + \ldots + (aa'_n) a_n;$$

so $\{a_1, \ldots, a_n\}$ generates the ideal A.

Thus every ideal of R is finitely generated, and hence R is Noetherian by Theorem 11.2.

Let $x = a/b$ ($a, b \in R$, $b \neq 0$) be an element of the field of fractions F of R which is integral over R. Then there is a natural number $n \geqslant 1$ and a family $(a_i)_{i \in [1, n]}$ of elements of R such that

$$x^n + a_1 x^{n-1} + \ldots + a_n = 0.$$

Let A be the R-submodule of F generated by $\{e, x, \ldots, x^{n-1}\}$. Then $b^{n-1}A \subseteq R$, and so A is a fractional ideal of R. Since $xA \subseteq A$ (cf. Theorem 13.1) it follows easily that $A^2 \subseteq A$; on the other hand, A is certainly included in A^2. So $A = A^2$ and hence, since every fractional ideal of a Dedekind ring is invertible, $A = R$. Thus, in particular, x belongs to R.

Hence R is integrally closed in its field of fractions.

We have already shown in Theorem 16.1 that every non-zero proper prime ideal of a Dedekind domain is maximal.

This completes the proof that (a) implies (b).

(2) Next we show that (b) implies (c). So let R be an integral domain with identity satisfying the three conditions of (b). Since $FI(R)$ is a semigroup with identity all that remains to be shown is that every non-zero fractional ideal of R is invertible in $FI(R)$.

The first move in this direction is to prove that every non-zero ideal includes a product of non-zero prime ideals. Suppose, to the contrary, that the set of non-zero ideals A such that A does not include a product of non-zero prime ideals is not empty. Then, since R is a Noetherian ring, this set has a maximal element, A_0 say. Clearly A_0 cannot itself be a prime ideal; so there exist elements a_1 and a_2 of R such that $a_1 \notin A_0$, $a_2 \notin A_0$, but $a_1 a_2 \in A_0$. Set $A_1 = A_0 + (a_1)$, $A_2 = A_0 + (a_2)$. Then $A_1 \supset A_0$, $A_2 \supset A_0$ and so it follows from the maximal property of A_0 that both A_1 and A_2 include products of non-zero prime ideals; hence so does their product $A_1 A_2$. But clearly $A_1 A_2 \subseteq A_0$; so A_0 includes a product of non-zero prime ideals. This is a contradiction; thus we have established that every non-zero ideal includes a product of non-zero prime ideals.

Next let P be any non-zero proper prime ideal of R, P^* its reciprocal. Since P is an integral ideal, P^* certainly includes R; we contend that this inclusion is proper. If a is any non-zero element of P, the principal ideal (a) includes a product of non-zero prime ideals; among all such finite families $(P_i)_{i \in I}$ of non-zero prime ideals such that $\prod_{i \in I} P_i \subseteq (a)$, choose one such that Card I is least, say

$$P_1 \ldots P_r \subseteq (a) \subseteq P.$$

Since P is prime, at least one of the ideals $P_1, \ldots, P_r$ is included in P, say $P_1 \subseteq P$. It follows that $P_1 = P$, since, by hypothesis, all prime ideals of R are maximal. Since $P_1 \ldots P_r$ is a shortest product of prime ideals included in (a), the product $P_2 \ldots P_r$ is not included in (a); so there is an element b of $P_2 \ldots P_r$ such that $b \notin (a)$, i.e. such that b/a does not belong to R. But if p is any element of $P = P_1$ we have

$$pb \in P_1(P_2 \ldots P_r) \subseteq (a)$$

and so $p(b/a) \in R$. Thus $b/a \in P^*$, and so $P^* \supset R$.

Our next contention is that every non-zero integral ideal of R is invertible in $FI(R)$. Suppose, to the contrary, that the set of non-zero integral ideals of R without inverses is non-empty; since R is Noetherian this set has a maximal element, A say. Then $A \neq R$, since R is invertible, and hence, according to Theorem 3.5, A is included in a maximal ideal, P say. We then have $P^* \subseteq A^*$ and so

$$A = AR \subseteq P^*A \subseteq A^*A \subseteq R.$$

We claim that in fact $A \subset P^*A$. Otherwise we have $xA \subseteq A$ for every element x of P^*. Since R is Noetherian, A is a finitely generated R-module; so it follows from Theorem 15.1 that every element x of P^* is integral over R and hence, since R is integrally closed, $x \in R$. So $P^* \subseteq R$, in contradiction to what we established in the previous paragraph. Thus we have $P^*A \supset A$, and consequently (since A is a maximal non-invertible ideal) P^*A is invertible, with inverse B say. Then $(BP^*)A = B(P^*A) = R$ and A is invertible, in contradiction to its definition. Thus we have established that every non-zero integral ideal of R is invertible.

Finally, let A be any non-zero fractional ideal of R; we may write $A = B_1 B_2^{-1}$ where B_1 and B_2 are integral ideals and B_2 is invertible. According to the preceding paragraph, B_1 is also invertible. Then clearly $B_2 B_1^{-1}$ is an inverse for A.

Hence $FI(R)$ is a group.

(3) Finally we show that (c) implies (a). So let R be an integral domain such that every non-zero fractional ideal is invertible.

At the very beginning of the proof of this theorem we showed that every invertible integral ideal is finitely generated. Thus, under the hypothesis (c), every integral ideal of R is finitely generated and so R is Noetherian (by Theorem 11.2).

We wish to show that every non-zero proper (integral) ideal of R has a factorisation as a product of prime ideals. Suppose this is not the case; then the set of non-zero proper ideals of R which do not have such a factorisation is non-empty and hence has a maximal element, A say. Theorem 3.5 shows that A is included in a maximal ideal P; this inclusion is of course proper, since P is prime and hence (trivially) has a prime factorisation. It follows that AP^{-1} is a proper integral ideal of R. Since $P \subset R$ we have $P^{-1} \supseteq R$ and hence $AP^{-1} \supseteq A$; but the inclusion here must be proper, since otherwise we would have $AP^{-1} = A$ and so (since A is invertible) $P = R$.

Thus $A \subset AP^{-1} \subset R$, i.e. AP^{-1} is a non-zero proper integral ideal of R properly including A. From the maximal property of A we deduce that AP^{-1} has a factorisation as a product of prime ideals, say

$$AP^{-1} = P_1 \dots P_k.$$

It follows at once that

$$A = PP_1 \dots P_k,$$

i.e. A has a factorisation as a product of prime ideals. This, however, contradicts the definition of A.

Hence R is a Dedekind domain.

This completes the proof.

We must now link up the discussion of Dedekind domains with the example of non-unique factorisation which we gave at the beginning of this section. The connexion is made by the following theorem, for the proof of which we refer the reader to Lang, *Algebraic numbers*.

THEOREM 16.3. *Let R be a Dedekind domain, F its field of fractions. If E is a finite algebraic extension of F then the relative integral closure of R in E is also a Dedekind domain.*

(For an explanation of the term 'finite algebraic extension' see Adamson, *Introduction to Field Theory*.)

In the theorem take $R = \mathbf{Z}$, the ring of integers; then $F = \mathbf{Q}$, the field of rational numbers. The field $\mathbf{Q}(i\sqrt{5})$ is a finite algebraic extension of $\mathbf{Q}$, and it is easy to check that the relative algebraic closure of $\mathbf{Z}$ in $\mathbf{Q}(i\sqrt{5})$ is the subring S consisting of elements of the form $a + b\sqrt{5}i$ where a and b are integers. Since $\mathbf{Z}$ is a Dedekind domain, it follows from the theorem that S is also a Dedekind domain. Thus, although the elements of S have non-unique factorisation into irreducible elements, the ideals of S have essentially unique factorisation into prime ideals.

EXERCISES 3

1. Show that the ring of Gaussian integers $\mathbf{G}$ is isomorphic to the residue class ring $P(\mathbf{Z})/I$ where $P(\mathbf{Z})$ is the ring of polynomials with integer coefficients and I is the principal ideal generated by the polynomial $X^2 + 1$.

2. Show that the principal ideals (3) and $(1 + i)$ are prime in $\mathbf{G}$ but that (2) is not.

3. Let p be a prime number; let $\mathbf{Q}_p$ be the set of rational numbers consisting of 0 and all non-zero rational numbers of the form ap^n where a and n are integers and p does not divide a. Prove that $\mathbf{Q}_p$ is a subring of $\mathbf{Q}$ and that the mapping v defined by setting $v(ap^n) = |a|$ for all non-zero elements ap^n of $\mathbf{Q}_p$ is a Euclidean norm on $\mathbf{Q}_p$.

4. Let R be a Euclidean ring. Prove that a left R-module V is injective if and only if $rV = V$ for every non-zero element r of R.

5. Let α be an epimorphism from a commutative ring R onto a ring S with kernel K; let I be an ideal of R which includes K. Show that $\alpha(I)$ is prime or maximal if and only if I is prime or maximal respectively.

6. Let R be a commutative ring, N the ideal of nilpotent elements of R. Show that every prime ideal of R includes N. Let a be a non-nilpotent element, A the set of powers of a, E the set of ideals I of R such that $I \cap A = \varnothing$; show that E has a maximal element M and that M is a prime ideal. Deduce that N is the intersection of all the prime ideals of R.

7. Let R, R' be commutative rings with identity elements $e_R, e_{R'}$, S, S' subsets of R, R' respectively and α a homomorphism from R to R' such that $\alpha(S) \subseteq S'$. Show that there exists a unique homomorphism α_* from $R[S^{-1}]$ to $R'[(S')^{-1}]$ such that $\alpha_*(a/e_R) = \alpha(a)/e_{R'}$ for all elements a of R.

8. Let R be a commutative principal ideal ring, S a multiplicative system in R. Show that $R[S^{-1}]$ is a principal ideal ring.

9. Let R be a commutative ring, P a prime ideal of R and S the complement of P in R. Show that $R[S^{-1}]$ has a unique maximal ideal consisting of all elements of the form p/s where $p \in P$ and $s \in S$.

10. Let R be an integrally closed integral domain, F its quotient field. If f and g are monic polynomials with coefficients in F such that fg has coefficients in R show that both f and g have coefficients in R.

11. Let R be a Dedekind domain, S any multiplicative system in R. Prove that $R[S^{-1}]$ is a Dedekind domain.

READING LIST

ADAMSON, I. T., *Introduction to field theory*, Edinburgh, 1964.

ADAMSON, I. T., *Rings, modules and algebras*, Edinburgh, 1972.

BOURBAKI, N., *Algèbre* (Ch. 2, Algèbre linéaire), Paris, 1962.

BOURBAKI, N., *Algèbre commutative* (Ch. 2, Localisation), Paris, 1961.

LAMBEK, J., *Lectures on rings and modules*, Waltham, Mass., 1966.

LANG, S., *Algebraic numbers*, Reading, Mass., 1964.

McCOY, N. H., *Rings and ideals*, Carus Mathematical Monograph No. 8, 1948.

McCOY, N. H., *The theory of rings*, New York, 1964.

NORTHCOTT, D. G., *Ideal theory*, Cambridge, 1953.

WEISS, E., *Algebraic number theory*, New York, 1963.

ZARISKI, O. and SAMUEL, P., *Commutative algebra*, Vol. 1, Princeton, N.J., 1958.

Abelian group, 7
Artinian module, 87
Artinian ring, 89
Ascending chain, 85
Associate, 94
Associative, 3
Automorphism
    definition, 2
    identity automorphism, 23

Balanced mapping, 78
Basis, 38
Bidual, 49

Chain, 85
Characteristic, 28
Closed, 1, 35
Commutative, 4
Compatible, 6
Composition
    external law of composition, 32
    internal law of composition, 1
Contracted ideal, 112
Coset, 40
Couniversal, 57

Descending chain, 85
Direct product, 55
Direct sum
    external direct sum, 58
    internal direct sum, 61
Direct summand, 63
Distributive, 6
Divisible abelian group, 74
Divisor, 8
Domain
    Dedekind domain, 122
    Euclidean domain, 102
    Factorisation domain, 95
    Integral domain, 9
    Principal ideal domain, 101
    Unique factorisation domain, 97
Dual basis, 50
Dual module, 49

Epimorphism
    Canonical epimorphism, 24, 42, 55
    Definition, 2
Exact sequence
    Definition, 44
    Short exact sequence, 45
    Split exact sequence, 63
Extension, 72, 112
Euclidean domain, 102
Euclidean norm, 102

Factor
    Definition, 94
    Highest common factor, 99, 100
    Proper factor, 95
Factorisation
    Definition, 95
    Factorisation domain, 95
    Unique factorisation domain, 97
Field, 9
Field of fractions, 110
Finitely generated, 19, 36
Fractional ideal, 122
Free module, 66
Free subset, 37

Generated, 17, 36, 108
Generating system, 19, 36
Group, 7

Homomorphism, 2, 23, 41

Ideal
    Definition, 16
    Fractional ideal, 122
    Integral ideal, 122
    Maximal ideal, 19
    Minimal ideal, 19
    Principal ideal, 17
    Proper ideal, 16
    Zero ideal, 16
Identity, 4
Identity automorphism, 23
Induced law of composition, 1, 6, 35
Induced homomorphism, 26, 50
Injective module, 72
Injective representation, 60

Integral closure, 118
Integral domain, 9
Integral element, 116
Integral ideal, 122
Integrally dependent, 116
Integrally closed, 118
Inverse, 5
Invertible, 5
Irreducible element, 95
Irreducible module, 36
Isomorphism, 2

Kernel, 24, 42

Law of composition
  external law, 32
  internal law, 1
Level off, 85
Linear combination, 36
Linear form, 49
Linear functional, 49
Linearly dependent, 37
Linearly independent, 37

Maximum condition, 86
Minimum condition, 86
Module
  Artinian module, 87
  Definition, 32, 33
  Factor module, 40
  Free module, 66, 67
  Injective module, 72
  Irreducible module, 36
  Noetherian module, 87
  Projective module, 69
  Projective module, 69
  Quotient module, 40
  Simple module, 36
Monomorphism
  Canonical injection monomorphism, 56
  Definition, 2
  Inclusion monomorphism, 23, 41
Multiple, 7, 8
Multiplicative system, 108

Neutral element, 4
Natural mapping, 66, 78
Noetherian module, 87
Noetherian ring, 89

Opposite ring, 14

Power, 7
Prime
  Prime element, 97
  Prime ideal, 105
  Prime to S, 113
  Relatively prime, 99
Product, 1, 2, 20
Projection epimorphism, 55
Projective module, 69
Projective representation, 58

Quasi-finite, 37
Quaternions, 12
Quotient module, 40
Quotient ring, 110

Radical, 107
Reciprocal, 123
Residue class, 22
Ring
  Commutative ring, 8
  Definition, 8
  Division ring, 9
  Opposite ring, 14
  Principal ideal ring, 101
  Quotient ring, 110
  Residue class ring, 22
  Ring of endomorphisms, 11
  Ring of fractions, 110
  Ring with identity, 8

S-inverting, 110
Scalar multiplication, 32
Semigroup, 7
Stable, 1, 35
Submodule, 35, 36
Subring, 14
Sum, 1, 2, 20, 37

Tensor product, 77, 82

Unique factorisation, 97, 123
Unit, 13
Universal, 59, 66, 79, 111

Vector space, 33

Zero element, 4
Zero homomorphism, 23, 31